The first twenty-five years of Superbrain

1984–2008

Diarmuid Early and Des MacHale

The United Kingdom Mathematics Trust

The first twenty-five years of Superbrain

© 2011 United Kingdom Mathematics Trust

All rights reserved. No part of this publication may be reproduced or transmitted in any form or by any means, electronic or mechanical, including photocopy, recording, or any information storage and retrieval system, without permission in writing from the publisher.

Published by The United Kingdom Mathematics Trust.

Maths Challenges Office, School of Mathematics, University of Leeds, Leeds, LS2 9JT, United Kingdom

http://www.ukmt.org.uk

First published 2009 by Highperception Ltd.
978-1-906338-01-5

This edition published 2014

ISBN 978-1-906001-22-3

Printed in the UK for the UKMT by Charlesworth Press, Wakefield.
http://www.charlesworth.com

Introduction

The Superbrain examination had its origins at University College Cork in 1983, when a discussion arose between Science and Arts mathematics students on the one hand and students of Engineering on the other hand as to who were the superior mathematicians. Engineering students claimed that because of their higher requirements for entry into university they were clearly superior, but honours students in Science and Arts hotly disputed this. So, a challenge went out.

This led to the organization of an annual competitive examination, open to all full-time registered students of UCC, undergraduate and postgraduate alike, and from all faculties. It was dubbed the "Superbrain Competition" and since 1984 the questions have been set and marked by Professor Des MacHale. The topics were relatively elementary so as not to discriminate against newly-arrived first year students, and included calculus, geometry, combinatorics, algebra, elementary number theory, and probability — but the questions were often disguised, and a good deal more difficult than contestants were used to. Since 1990, the Superbrain has been used to select the UCC team for the Irish Mathematics Intervarsity Competition, with the top four students forming the UCC team. UCC has had outstanding success in this competition, winning it more often than any other university.

In this book, Des MacHale and Diarmuid Early, three-times winner of the Superbrain and twice individual winner of the Irish Mathematics Intervarsity, reveal the complete solutions of the Superbrain for the first time in twenty-five years.

In presenting solutions to problems, we have attempted to show how the result can be established, but also, wherever possible, to demonstrate the motivation that might lead the reader to discover the solution. We hope that this book will be useful to all of those preparing for Olympiad Examinations and other competitive examinations in mathematics, to teachers of mathematics, and to all of those who enjoy solving problems in mathematics.

Unlike many Mathematical Olympiads, competitors are allowed to use calculators in the Superbrain, though this is rarely necessary. Full explanations and proofs are required. The competition is open to all full-time students of University College Cork. The time limit is 3 hours, and the 10 questions are each marked out of 10. The winning score is often around 70%, but marks of over 90% have been achieved.

The pass mark is 0%. The papers have been constructed by Des MacHale, and about half of the problems are original. The others are heavily disguised versions of problems that he has come across.

The authors would like to thank a number of people for composing or suggesting unexpected and elegant solutions some of the problems, including Stephen Casey, Prithwijit De, Patrick Fitzpatrick, Peter Hegarty, Hugh McManus, Cillian McNamara, Sean Murphy, Eoin Parker, Christian Van Den Bosch, Michael White, and all those who have sat the Superbrain exam over the years. Particular thanks are owed to Finbarr Holland, for his invaluable help in offering alternative solutions to various problems and in correcting errors in the text. Any remaining errors, each of the authors blames on the other.

We thank Michael Green of the *Information about Ireland* site, for permission to use a map of the counties of Ireland.

Publisher's Remarks

This is an extraordinary collection of mathematical problems, laced with some puzzles. People who enjoy the problems of mathematical olympiads will certainly enjoy *Superbrain*. However, the menu of problems is more extensive than the usual diet of algebra, combinatorics, geometry and number theory, although those topics are well represented.

There are also ingenious calculus problems, some of which stray towards elementary analysis. On the other hand, there are some very entertaining and witty "what is the next term of this sequence" questions which could be answered by a sufficiently clever young child. Problem 2 of 1986 is an example of such a well constructed question. Every member of a family could enjoy that, without the need to be a mathematics specialist.

HP$^\text{n}$ 2009

Past Winners

Year	Superbrain	Winner
2008	25th	Stephen Casey
2007	24th	Diarmuid Early
2006	23rd	Stephen Casey
2005	22nd	Diarmuid Early
2004	21st	Diarmuid Early
2003	20th	Jerry Buckley
2002	19th	Thomas Cooney
2001	18th	Eamonn Long
2000	17th	Thomas Cooney
1999	16th	Deirdre O'Brien
1998	15th	John Sullivan
1997	14th	Hugh McManus
1996	13th	Damien Fennell
1995	12th	John Sullivan
1994	11th	Peter Hegarty
1993	10th	Peter Hegarty
1992	9th	Peter Hegarty
1991	8th	Liam Kelleher
1990	7th	Peter Hegarty
1989	6th	John Flynn
1988	5th	Jimmy Scully
1987	4th	John Flynn
1986	3rd	John O'Connell
1985	2nd	James Cunnane
1984	1st	Stephen Buckley

Problems

First Superbrain, 1984	2
Second Superbrain, 1985	3
Third Superbrain, 1986	5
Fourth Superbrain, 1987	6
Fifth Superbrain, 1988	7
Sixth Superbrain, 1989	8
Seventh Superbrain, 1990	9
Eighth Superbrain, 1991	10
Ninth Superbrain, 1992	11
Tenth Superbrain, 1993	12
Eleventh Superbrain, 1994	13
Twelfth Superbrain, 1995	14
Thirteenth Superbrain, 1996	16
Fourteenth Superbrain, 1997	17
Fifteenth Superbrain, 1998	18
Sixteenth Superbrain, 1999	19
Seventeenth Superbrain, 2000	20
Eighteenth Superbrain, 2001	21
Nineteenth Superbrain, 2002	22
Twentieth Superbrain, 2003	23
Twenty-First Superbrain, 2004	24
Twenty-Second Superbrain, 2005	26
Twenty-Third Superbrain, 2006	28
Twenty-Fourth Superbrain, 2007	30
Twenty-Fifth Superbrain, 2008	31

Solutions

First Superbrain, 1984 34
Second Superbrain, 1985 42
Third Superbrain, 1986 50
Fourth Superbrain, 1987 56
Fifth Superbrain, 1988 63
Sixth Superbrain, 1989 69
Seventh Superbrain, 1990 77
Eighth Superbrain, 1991 86
Ninth Superbrain, 1992 94
Tenth Superbrain, 1993 103
Eleventh Superbrain, 1994 110
Twelfth Superbrain, 1995 117
Thirteenth Superbrain, 1996 123
Fourteenth Superbrain, 1997 130
Fifteenth Superbrain, 1998 139
Sixteenth Superbrain, 1999 147
Seventeenth Superbrain, 2000 154
Eighteenth Superbrain, 2001 162
Nineteenth Superbrain, 2002 170
Twentieth Superbrain, 2003 178
Twenty-First Superbrain, 2004 185
Twenty-Second Superbrain, 2005 192
Twenty-Third Superbrain, 2006 201
Twenty-Fourth Superbrain, 2007 207
Twenty-Fifth Superbrain, 2008 217

Part 1:

Problems

First Superbrain, 1984 — Problems

1. Using each number exactly once, place the numbers 1, 2, 3, 4, 5, 6, 7, 8 and 9 in a 3 × 3 square so that all rows, columns and diagonals sum to different totals.

2. If x and y are positive integers, find all solutions of the equation
$$2xy - 4x^2 + 12x - 5y = 11.$$

3. If ABC is an acute-angled triangle, show with proof how to find points S and R on the line BC, P on the line AB, and Q on the line AC, such that $PQRS$ is a square.

4. Ten books are arranged in a row on a shelf. How many ways can this be done if one particular book A must always be to the left, but not necessarily immediately to the left, of another book B?

5. Assuming that
$$\lim_{x \to 0} \sqrt[x]{\frac{1+x}{1-x}}$$
exists, find its value.

6. If $n = 2^k$ for $k \geq 1$, show that $\binom{n}{r}$, the number of combinations of n things r at a time, is an even number for $0 < r < n$.

7. If $A = (a, b, c, d)$ is a set of four distinct elements, is it possible to define a closed binary operation $*$ on A such that the associative law $x * (y * z) = (x * y) * z$ never holds for any triple $x, y, z \in A$, equal or distinct?

8. Let $\alpha = 1 + \frac{x^3}{3!} + \frac{x^6}{6!} + \ldots$, $\beta = x + \frac{x^4}{4!} + \frac{x^7}{7!} + \ldots$ and $\gamma = \frac{x^2}{2!} + \frac{x^5}{5!} + \frac{x^8}{8!} + \ldots$. Assuming that all three series converge for $x \in \mathbb{R}$, prove that
$$\alpha^3 + \beta^3 + \gamma^3 = 1 + 3\alpha\beta\gamma.$$

9. Prove that $\cos 29°$ is not a rational number.

10. Form a nine digit number using each of the digits 1, 2, 3, 4, 5, 6, 7, 8, 9, once and only once so that the number formed by the nine digits is a multiple of nine, the number formed by the first eight digits is a multiple of eight, the number formed by the first seven is a multiple of seven and so on, i.e. the number formed by the first n digits is a multiple of n, for $1 \leq n \leq 9$.

Second Superbrain, 1985 — Problems

1. In a darts competition, each dart scores 40, 39, 24, 23, 17 or 16 points. How many darts must be thrown to get exactly 100 points?

2. Five points lie inside an equilateral triangle of side 2 units. Prove that at least two of the points are no more than a unit distance apart.

3. Find all prime numbers that can be written in the form $a^4 + 4b^4$, where a and b are positive integers.

4. If A, B and C are angles with
$$\sin A + \sin B + \sin C = 0 = \cos A + \cos B + \cos C,$$
prove that
$$\cos 3A + \cos 3B + \cos 3C = 3\cos(A + B + C).$$

5. In how many different ways is it possible to pay a punt (100p) using 50p, 10p and 5p pieces only?

6. If, in the accompanying figure, $ABDC$, $BEFD$, $EGHF$ are squares, find with proof, the size of angle $\alpha + \beta$.

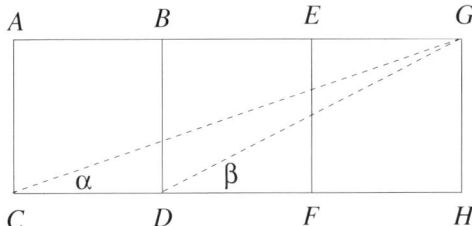

7. Given a triangle ABC, show how to find a point P such that
$$|PA|^2 + |PB|^2 + |PC|^2$$
is as small as possible.

8. What is the maximum and the minimum number of Friday the thirteenths that can occur in any calendar year?

9. By considering
$$\int_0^1 \frac{x^4(1-x)^4}{1+x^2}\,dx,$$
show that $\pi < \frac{22}{7}$.

10. In this long division each asterisk stands for a whole number. Reconstruct all the calculations, given that there is no remainder.

```
                    *  7  *  *
     *  *  * | *  *  *  *  *  *  *  *
               *  *  *  *
               ─────────
                     *  *  *
                     *  *  *
                     ────────
                        *  *  *  *
                        *  *  *
                        ─────────
                           *  *  *  *
                           *  *  *  *
                           ─────────
```

Third Superbrain, 1986 — Problems

1. 1986 is the year Cork 801. Show how to write the number 801 using each of the digits 0, 1, 2, 3, 4, 5, 6, 7, 8, 9 once and only once, using any mathematical operation you wish.

2. What, in your opinion is the next term of the sequence, and why?
$$F, 4, E, S, 9, S, E, 5, E, \ldots$$

3. Find the value of $\cos^2 0° + \cos^2 1° + \ldots + \cos^2 89°$ correct to twenty decimal places.

4. Which of the following numbers is bigger, and why?
$$\int_0^1 \sqrt[4]{1-x^7}\,dx \quad \text{or} \quad \int_0^1 \sqrt[7]{1-x^4}\,dx$$

5. Find the value of
$$\sum_{n=1}^{\infty} \frac{2n+1}{n^4 + 2n^3 + n^2}.$$

6. A thin rod is broken into three pieces. What is the probability that a triangle can be formed from the three pieces?

7. Prove that
$$\sum_{r=1}^{n} \frac{1}{r}$$
is not an integer for $n > 1$.

8. A right-angled triangle has all of its sides an integer length. If the length of the perimeter equals the area, find all such triangles.

9. Is it possible to have a right-circular cone in which the total surface area is equal to the volume? Justify your answer.

10. Find the smallest positive integer that can be expressed as the sum of two squares of positive integers in two different ways.

Fourth Superbrain, 1987 — Problems

1. What, in your opinion, is the next term in the sequence
$$3, 5, 1, 15, 11, 10,$$
and why?

2. What is the minimum number of different calendars it is necessary to have to cover every year? Justify your answer.

3. Decide whether or not the number
$$314154314155314156314157314158314159$$
is the square of an integer.

4. Evaluate
$$\sum_{n=1}^{\infty} \tan^{-1}\left(\frac{1}{n^2 + n + 1}\right).$$

5. If a, b, c, d, x, y, z and w are positive integers with $\frac{a}{b} > \frac{c}{d}$ and $\frac{x}{y} > \frac{z}{w}$, prove or disprove that
$$\frac{a + x}{b + y} > \frac{c + z}{d + w}.$$

6. Evaluate the definite integral
$$\int_0^\pi \frac{x \sin x}{1 + \cos^2 x} dx.$$

7. What are the minimum and maximum values of the function
$$f(x) = \frac{x^3}{3} + \frac{x^2}{2} - 2x$$
on the interval $[-3, 4]$?

8. A series of numbers, beginning with 17, is in both arithmetic and geometric progression. Find the sum of the first million terms of the series.

9. ABC is a triangle. Show how to construct a square $XYZW$ such that A lies on the line XY, B lies on the line YZ and C lies on the line XW.

10. Find all integers x and y which satisfy the equation $x^2 - 3y^2 = 17$.

Fifth Superbrain, 1988 — Problems

1. What, in your opinion is the next term in the sequence F, G, J, L, P, Q, and why?

2. Find all real numbers a, b and c such that
$$a + b + c = 0 = \frac{1}{a} + \frac{1}{b} + \frac{1}{c}.$$

3. If n is a natural number, prove that
$$n^n \geq 1.3.5.7\ldots(2n-1).$$

4. Find all real numbers x, y such that $(x + iy)^2 = -i$, where $i^2 = -1$.

5. Evaluate
$$\int \frac{\sqrt{x}}{1 + \sqrt[3]{x}} dx.$$

6. Investigate
$$\lim_{x \to 0} \frac{\sin \frac{1}{x}}{\sin \frac{1}{x}}.$$

7. Find the value of $\dfrac{1}{3} + \dfrac{2}{5} + \dfrac{4}{17} + \dfrac{8}{257} + \ldots = \displaystyle\sum_{n=0}^{\infty} \dfrac{2^n}{2^{2^n} + 1}.$

8. A plane curve is given by
$$y = ax^3 + bx^2 + cx + d,$$
for $a \neq 0$. Find conditions on a, b, c and d which ensure that the curve has a maximum point and a minimum point locally. Under these conditions show that the curve has a point of inflection that lies midway between the local maximum point and the local minimum point.

9. A circle is inscribed in a quadrant of a circle whose radius is 10cm. Find the radius of the inscribed circle.

10. Prove, using vectors, that the angle in a semi-circle is a right angle.

Sixth Superbrain, 1989 — Problems

1. Given the coordinates of three distinct points A, B, C in the plane, give ten essentially different methods of deciding whether or not A, B, and C lie on the same line.

2. Given n consecutive positive integers, show that $n!$ is a factor of their product.

3. Evaluate the definite integral
$$\int_0^{\frac{\pi}{2}} \frac{\sin^{17} \theta}{\cos^{17} \theta + \sin^{17} \theta} d\theta.$$

4. Find any three positive whole numbers a, b, and c which satisfy the equation $a^3 + b^4 = c^5$.

5. Given five distinct real numbers x_1, x_2, x_3, x_4, x_5, prove that at least two of them satisfy the inequality
$$0 < \frac{x_i - x_j}{x_i x_j + 1} < 1.$$

6. Let $ABCD$ be a trapezium with AB parallel to DC. The diagonals AC and BD intersect at R and the line MRN is parallel to AB, where M lies on AD and N lies on BC. Prove that $|MR| = |RN|$.

7. Evaluate $\dfrac{1}{3} + \dfrac{1}{3} + \dfrac{1}{5} + \dfrac{4}{45} + \ldots = \displaystyle\sum_{n=1}^{\infty} \dfrac{n 2^n}{(n+2)!}$.

8. The lengths of the sides of a triangle are in geometric progression with common ratio r. Prove that $\frac{1}{\phi} < r < \phi$, where $\phi = \frac{1+\sqrt{5}}{2}$.

9. If a, b, c are numbers such that
$$\frac{1}{a} + \frac{1}{b} + \frac{1}{c} = \frac{1}{a+b+c},$$
prove that
$$\frac{1}{a^{23}} + \frac{1}{b^{23}} + \frac{1}{c^{23}} = \frac{1}{a^{23} + b^{23} + c^{23}}.$$

10. Each side of a right-angled triangle has an integer length and the length of the perimeter is twice the area of the triangle. Find the length of each side.

Seventh Superbrain, 1990 — Problems

1. A car is driven from A to B at an average speed of $40km$ per hour and is then driven from B to A along the same path at an average speed of $60km$ per hour. What is the average speed of the car for the entire journey?

2. If n, x, y and z are all positive integers, find all solutions of the equation
$$n^x + n^y = n^z.$$

3. Prove that
$$\sin 18° = \frac{\sqrt{5} - 1}{2}.$$

4. At what time between 1pm and 2pm do the minute hand and the hour hand of a clock coincide exactly?

5. Solve the equation
$$30x^4 + 61x^3 + 90x^2 + 61x + 30 = 0.$$

6. A calendar month is said to be "bad" if it has five Mondays. What is the maximum and minimum number of "bad" months in a year?

7. Evaluate
$$\sum_{r=1}^{n} r^4.$$

8. Prove that the distance of the point whose coordinates are (x_1, y_1) from the line whose equation is $ax + by + c = 0$ is
$$\left| \frac{ax_1 + by_1 + c}{\sqrt{a^2 + b^2}} \right|.$$

9. Evaluate
$$\int_0^1 \sqrt{x - x^2}\, dx.$$

10. If H is the orthocentre of a triangle ABC and O is the circumcentre, prove that
$$\vec{OA} + \vec{OB} + \vec{OC} = \vec{OH}$$
and
$$\vec{HA} + \vec{HB} + \vec{HC} = 2\vec{HO}.$$

Eighth Superbrain, 1991 — Problems

1. Show that no term of the sequence
$$11, 111, 1111, 11111, \ldots$$
is the square of an integer.

2. Find the 20th derivative, $\dfrac{d^{20}y}{dx^{20}}$, of the function
$$y = \frac{1}{1-x^2}.$$

3. The lengths of the sides of a given quadrilateral are a, b, c and d (labelled clockwise) and its area is A. Prove that $4A \leq (a+c)(b+d)$, with equality if and only if the figure is a rectangle.

4. Evaluate
$$\lim_{n \to \infty} \left[\sqrt{(n+4)(n+5)} - \sqrt{(n+2)(n+3)} \right].$$

5. Find the nth term of the sequence
$$\sum_{r=1}^{\infty} \frac{1}{r(r+1)}, \quad \sum_{r=1}^{\infty} \frac{1}{r(r+1)(r+2)}, \quad \sum_{r=1}^{\infty} \frac{1}{r(r+1)(r+2)(r+3)} \cdots$$

6. Solve the equations
   (a) $\dfrac{2x+1}{(x+3)(x-2)} + \dfrac{5x}{(2x+1)(x+3)} = \dfrac{x+3}{(2x+1)(x-2)}.$
   (b) $3\left(\sin^4\theta + \cos^4\theta\right) - 2\left(\sin^6\theta + \cos^6\theta\right) = 1.$

7. If $x+y+z = 0$, prove that
$$7(x^2+y^2+z^2)(x^5+y^5+z^5) = 10(x^7+y^7+z^7).$$

8. Let m, n and k be positive integers. One solution of the equation $(m!)(n!) = k!$ is $m = 6$, $n = 7$, $k = 10$. Find another solution with m, n and k all greater than 10.

9. If $f(x) = \frac{x}{2} + \cos x$ and $x \in [0, 2\pi]$, what are the greatest and least values of $f(x)$?

10. Prove that $4^{50} + 2^{50} + 1$ is not a prime number.

Ninth Superbrain, 1992 — Problems

1. What, in your opinion, is the next term in each of the following sequences and why?
   (a) T, Q, P, H, H, O, N?
   (b) 2, 3, 4, 5, 7, 8, 9, 11, 13, 17?
   (c) 1, 2, 6, 2, 1, 7, 5, 4, 3, 3, 3, 4, 6?
   (d) 1827, 6412, 5216, 3435, 1272, 9100?
   (e) 1, 3, 11, 25, 137, 49, 363, 761?

2. Let $p_1 < p_2 < p_3$ be prime numbers such that $p_1^2 + p_2^2 + p_3^2$ is also a prime number. Find the value of p_1.

3. Find any triangle ABC such that each side has an integer length, and one angle of the triangle is twice the size of another angle of the triangle.

4. Evaluate
$$\sum_{r=1}^{\infty} \tan^{-1}\left(\frac{1}{2r^2}\right).$$

5. Solve the equation
$$(x-1)(x-2)(x-3)(x-4) + 9 = 0.$$

6. Find the maximum value of the function
$$f(\theta) = \sin^3\theta - 4\sin\theta.$$

7. Let P_i be distinct points with coordinates (x_i, y_i) for $i \in [1, n]$. Find the coordinates of a point P such that
$$\sum_{i=1}^{n} |PP_i|^2$$
is as small as possible.

8. Evaluate
$$\int \frac{dx}{\sin^4 x \cos^4 x}.$$

9. If a, b, c and d are real numbers, prove that $a^4 + b^4 + c^4 + d^4 \geq 4abcd$, assuming only that $x^2 \geq 0$ for any real number x.

10. If m and n are odd positive integers such that m^m divides n^n, prove or disprove that m divides n.

Tenth Superbrain, 1993 — Problems

1. The length of each side of a rectangle is a whole number, and the area of the rectangle measured in square metres is equal to the length of the perimeter of the rectangle measured in metres. If the rectangle is not a square, find with proof the dimensions of the rectangle.

2. Evaluate
$$\sum_{r=1}^{n}(r^2+1)r!.$$

3. Find the lengths of the sides of any triangle in which each of the numbers $\tan A$, $\tan B$ and $\tan C$ is an integer.

4. Let S be a set with precisely n elements, where n is a fixed natural number. If a non-empty subset T of S is chosen at random, show that the probability that T has an odd number of elements is greater than the probability that T has an even number of elements.

5. If A and B are the midpoints of adjacent faces of a cubical box and X is a vertex common to both faces, find, with proof, the size of $\angle AXB$.

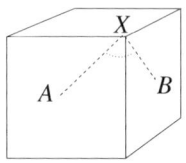

6. Find, with proof, all solutions of the equation $x! + y! = z!$, where x, y and z are all positive integers.

7. Find all angles A and B such that $\sin(A+B) = \sin A + \sin B$.

8. 131, 181, 15451 are all palindromic primes, i.e. primes equal to the number obtained when the digits in the number are reversed. Find all palindromic primes with an even number of digits.

9. The tangent to the curve $y(1+x^2) = 2$ at the point $P = \left(2, \frac{2}{5}\right)$ meets the curve again at Q. Find the coordinates of Q.

10. What, in your opinion, is the next term in the sequence 3, 11, 37, 101, 41, 7, 239,..., and why?

Eleventh Superbrain, 1994 — Problems

1. Given a square in the plane, show how to cut it into four pieces so that these pieces can be reassembled to form two squares of different sizes.

2. If n is a positive integer, show that $n(n+1)(n+2)$ is not the cube of an integer.

3. Evaluate
$$\sum_{n=1}^{\infty} \frac{n-1}{2^{n+1}}.$$

4. The sum of a number of positive integers is 100. Find the greatest value their product can have.

5. Let y be a function of x and x a function of y, and let $y_0 = y(x_0)$ for some x_0. If $\left.\frac{d^2y}{dx^2}\right|_{x_0} = 0$, find $\left.\frac{d^2x}{dy^2}\right|_{y_0}$.

6. If you were given a sequence which began 1, 10, 19, 82, 148, 187, 208, 325, 346, 565, what would you choose as the next term, and why?

7. Solve the equations $x^2 - yz = 1$, $y^2 - zx = 2$, $z^2 - xy = 3$ and verify your solutions.

8. Evaluate
$$\int \frac{dx}{e^x + 1}.$$

9. For any set S with exactly n elements, let ^nJ_r be the total number of subsets of S which contain *at least* r elements. Prove that
$$^nJ_r + {}^nJ_{r-1} = {}^{n+1}J_r \text{ for } 0 \leq r \leq n.$$

10. In an acute-angled triangle ABC, prove that
$$\tan A \tan B \tan C \geq 3\sqrt{3}.$$

Twelfth Superbrain, 1995 — Problems

1. In a hundred metres race, runner A, B and C all run at uniform speeds. If A beats B by 10 metres, and B beats C by 10 metres, by how many metres does A beat C?

2. If p, q, r and s are positive real numbers, prove that
$$\left(\frac{p^2+1}{q}\right)\left(\frac{q^2+1}{r}\right)\left(\frac{r^2+1}{s}\right)\left(\frac{s^2+1}{p}\right) \geq 16.$$

3. Evaluate
$$\int \frac{\sin x}{\sin x + \cos x}\,dx.$$

4. Let $f(x) = ax^3 + bx^2 + cx + d$ be a function $\mathbb{R} \to \mathbb{R}$, where a, b, c and d are fixed real numbers with $a > 0$. Find necessary and sufficient conditions for the inverse function $f^{-1} : \mathbb{R} \to \mathbb{R}$ to exist.

5. Let $2, 3, 5, 7, 11, \ldots, p_n, \ldots$ be the sequence of prime numbers. Show that
$$[2.3.5.7.11\ldots p_n] + 1$$
is never a square number.

6. Evaluate
$$\sum_{n=2}^{\infty} \log_2\left(1 - \frac{1}{n^2}\right).$$

7. A lampshade is in the shape of a frustrum of a right-circular cone with r the radius of the top, R the radius of the base and l the slant height. Find a formula for the area of the curved surface of the lampshade.

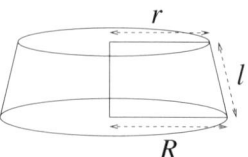

8. Without calculating either number, decide which of the numbers π^e or e^π is the greater.

[You may assume that $\pi > e$, where π is the ratio of the length of the circumference of a circle to its diameter, and e is the base of natural logarithms.]

9. A plane quadrilateral has sides of length a, b, c and d. If
$$a^2 + b^2 + c^2 + d^2 = ab + bc + cd + da,$$
prove that the diagonals of the figure are perpendicular to each other.

10. Let ABC be a triangle and r be the radius of the circle which touches all three sides of the triangle internally. Let p_1, p_2 and p_3 be the perpendicular distances of each of the vertices of the triangle to the opposite side (produced if necessary). Show that
$$\frac{1}{r} = \frac{1}{p_1} + \frac{1}{p_2} + \frac{1}{p_3}.$$

Thirteenth Superbrain, 1996 — Problems

1. Which is a closer fit, a square peg in a round hole or a round peg in a square hole?

2. Using coins with face value 1p, 2p, 5p, 10p, 20p and 50p, and these only, what is the largest amount of money that you could have and not be able to make up £1 (100p) exactly?

3. If $x = \sqrt{2} + \sqrt{3}$, evaluate
$$x^8 - 98x^4 + x + 17.$$

4. The plane is to be tiled with regular n-sided polygons, all of the same shape and size. Find all possible values that n can have.

5. Evaluate
$$\sum_{r=1}^{n} \sin rx.$$

6. If x, y and z are positive integers with $x \leq y \leq z$, find all solutions of the equation
$$xyz = x + y + z.$$

7. Both 30716 and 65419378 are natural numbers with the property that no digit is repeated. How many such natural numbers are there?

8. Evaluate
$$\int \frac{dx}{\sin x + \cos x}.$$

9. Find functions $f(x)$ and $g(x)$ such that
$$\frac{d}{dx}(f(x)g(x)) = \frac{df(x)}{dx}\frac{dg(x)}{dx},$$
where neither $f(x)$ nor $g(x)$ is a constant function.

10. Show that the sum of the squares of four consecutive numbers is never a square number.

Fourteenth Superbrain, 1997 — Problems

1. Find all prime numbers p such that $2p - 1$ and $2p + 1$ are also prime numbers.

2. A piece of wire 100cm in length is bent into the shape of a sector of a circle. Find the maximum value that the area A of the sector can have.

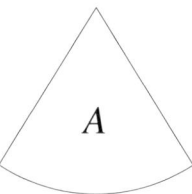

3. A tangent touches the curve $y = x^2(x-1)(x-3)$ at two distinct points. Find the equation of the tangent.

4. Show that there are exactly three kinds of plane triangle where the number of degrees in each angle is an integer divisor of 180.

5. If a and b are positive real numbers with $a > b$, prove that
$$\frac{(a-b)^2}{8a} < \frac{a+b}{2} - \sqrt{ab} < \frac{(a-b)^2}{8b}.$$

6. Evaluate
$$\int_0^\infty \frac{dx}{x^3 + 1}.$$

7. In the given triangle, the areas are as shown. Find the area x of the quadrilateral.

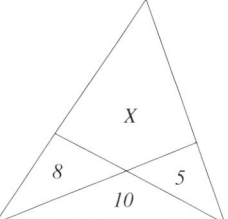

8. Evaluate
$$\sum_{n=1}^\infty \frac{n+2}{n(n+1)2^n}.$$

9. Show that there do not exist real numbers a, b with $\frac{1}{a} + \frac{1}{b} = \frac{1}{a+b}$, but that there do exist complex numbers with this property.

10. Using each of these ten digits 0, 1, 2, 3, 4, 5, 6, 7, 8, 9 once and only once, construct two five digit numbers with the largest possible product.

Fifteenth Superbrain, 1998 — Problems

1. The ages of Ann and Bill add up to 93. Ann is now three times as old as Bill was at the time when Ann was twice as old as Bill is now. How old is Bill now?

2. A man drives a car up a hill half a mile long at 30 miles per hour. How fast must he drive the half mile back down the same hill to average 60 miles per hour for the whole journey?

3. Let $ABCD$ be a plane quadrilateral, all angles of which are less than $180°$. Let a, b, c and d be the lengths of the sides and x and y the lengths of the diagonals. Show that $2(x+y) > a+b+c+d > x+y$.

4. Let a, b, c, x, y and z all be integers. Prove or disprove that there exist integers p, q and r with
$$(a^2 + b^2 + c^2)(x^2 + y^2 + z^2) = p^2 + q^2 + r^2.$$

5. A bag contains 36 snooker balls; some are black and the remainder are white, but there are more black than white. Two balls are drawn from the bag at the same time. If it is equally likely that the balls will be of the same colour as of different colours, how many black balls are in the bag?

6. A circular disc is divided into three pieces, A, B and C of equal area as in the diagram. If $0° < \alpha < 180°$, find an integer n such that $n° < \alpha < (n+1)°$.

7. Given a sequence of positive real numbers $a_1, a_2, a_3, \ldots$ with $a_1 = 1$ and $\sum_{i=1}^{n} a_i \leq a_{n+1}$, what is the largest value that $\sum_{i=1}^{\infty} \frac{1}{a_i}$ can have?

8. Evaluate $\displaystyle\int \frac{\sin^2 x}{\sin^2 x + 1} dx$.

9. Evaluate $\displaystyle\sum_{i=1}^{n} 2^i \tan\left(2^i \alpha\right)$ for all values of α for which this sum is defined.

10. Solve the equation $x^4 + 6x^3 + 11x^2 + 6x - 8 = 0$.

Sixteenth Superbrain, 1999 — Problems

1. A man whose birthday was January 1st died on February 29th, 1992. He was x years of age in the year x^2 AD. In what year was he born?

2. In the diagram, a square is divided into three regions of equal area A. If $0° < \alpha < 90°$, find α to the nearest degree.

   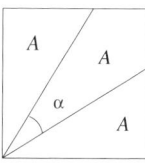

3. If p and $p^2 + 14$ are both prime numbers, find all possible values of p.

4. If a, b and c are positive integers with $\dfrac{1}{a} + \dfrac{1}{b} + \dfrac{1}{c} < 1$, find the maximum possible value of
$$\frac{1}{a} + \frac{1}{b} + \frac{1}{c}.$$

5. Evaluate
$$\int \frac{\sin\theta + \cos\theta}{3\sin\theta + 2\cos\theta}\,d\theta.$$

6. Let ABC be a triangle, and P be a point which lies inside it. Let AP meet BC at X, BP meet CA at Y and CP meet AB at Z.
   If $\dfrac{|AP|}{|PX|} = \dfrac{|BP|}{|PY|} = \dfrac{|CP|}{|PZ|} = k$, find the value of k.

   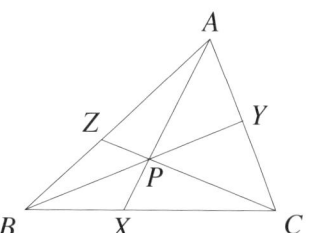

7. Evaluate $\displaystyle\sum_{n=1}^{\infty} \frac{1}{n\sqrt{n+1} + (n+1)\sqrt{n}}.$

8. Show that 128 is not the sum of distinct squares of integers.

9. If $\alpha_n = \dfrac{x^n}{y^n} + \dfrac{y^n}{x^n}$, express $\alpha_5 + \alpha_6$ in terms of α_1 only.

10. A and B are towns 20km and 30km from a straight stretch of river 100km long. Water is pumped from a point P on the river by pipelines to both towns. Where should P be located to minimise the total length of pipe $AP + PB$?

    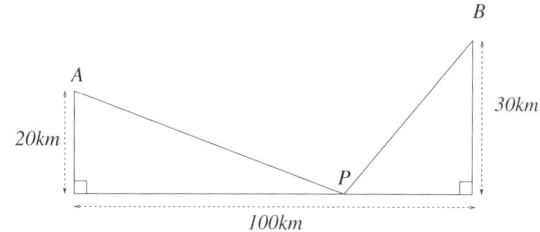

Seventeenth Superbrain, 2000 — Problems

1. Four horses A, B, C and D take part in a race. Including dead heats and non-finishers, how many different results are possible?

2. If x, y and z are positive integers, prove or disprove that there exist integers p, q and r such that $(x^2 + y^2 + z^2)^2 = p^2 + q^2 + r^2$.

3. (a) The usual construction for bisecting an angle requires three uses of the compasses. Show how to bisect and angle with just two uses of the compasses.

   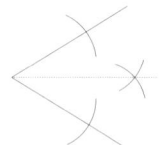

   (b) A circle is a set of points all in the plane which a given distance, known as the radius of the circle, from a given point, known as the centre of the circle. Show that a circle has only one centre.

4. An equilateral triangle is dissected into three equal triangles of area A as in the diagram. Find the measure of the angle θ to the nearest degree.

   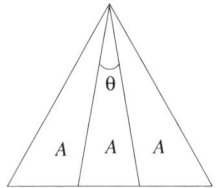

5. Evaluate
$$\int_0^{\frac{\pi}{2}} \frac{\sin^{2000}\theta}{\sin^{2000}\theta + \cos^{2000}\theta}\,d\theta.$$

6. Find all positive integers n such that $n^5 + n + 1$ is a prime number.

7. If $x = 4t^5 - t^3 + 1$ and $y = 5t^4 + t^2 - t$, find $\dfrac{d^2y}{dx^2}$ in terms of t only.

8. Factorise $27x^2 - 15xy - 112y^2$.

9. Evaluate $\sum_{r=0}^{n} \binom{n}{r}^2$.

10. Four cards, $A1$, $B2$, $C3$ and $D4$ are placed as in the diagram. Place twelve other cards $A2$, $A3$, $A4$, $B1$, $B3$, $B4$, $C1$, $C2$, $C4$, $D1$, $D2$ and $D3$, one in each remaining box, in such a way that no letter or number is repeated in any row, column or (length four) diagonal.

A1	B2	C3	D4

20

Eighteenth Superbrain, 2001 — Problems

1. The numbers 123456789 and 246913578 are two examples of nine-digit numbers in which no digit is repeated (we do not include zero as a digit). Find the sum of all such numbers.

2. If a, b and c are positive integers which satisfy $a^2 + b^2 = c^2$, show that 60 divides abc.

3. Evaluate $\sum_{n=1}^{\infty} \dfrac{n}{n^4 + n^2 + 1}$.

4. Let AOB be an angle and P a given point. Show how to find a point X on OB and a point Y on OA such that P is the midpoint of XY.

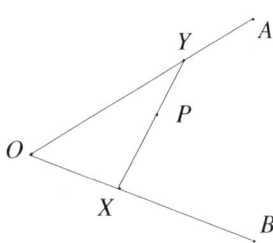

5. Express the sixth derivative of the function $y = \tan x$ as a polynomial in $\tan x$.

6. Evaluate $\displaystyle\int \dfrac{dx}{\sqrt{x} + \sqrt[3]{x}}$.

7. Starting with the first row, place a letter A, B, C, D, E or F in each square so that no letter is repeated in any row, column or length six diagonal.

8. Prove that $\sum_{r=1}^{n} \dfrac{1}{r^2} < 2$ for each natural number n.

9. Find any positive integer values of x, y, z and w which satisfy the equation
$$\dfrac{1}{x^3} + \dfrac{1}{y^3} + \dfrac{1}{z^3} = \dfrac{1}{w^3}.$$

10. The radius r and height h of a right-circular cone are both an integer number of centimetres, and the volume of the cone in cubic centimetres is equal to the total surface area of the cone in square centimetres. Find the values of r and h.

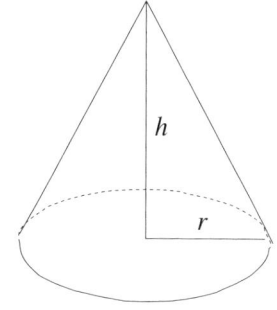

Nineteenth Superbrain, 2002 — Problems

1. (a) A leaden sphere of radius 1 metre is melted down and recast as a million spheres of lead shot, each of the same radius. Find the percentage increase in the surface area of the lead exposed.

   (b) To what occupation might questions such as 1(a) be of particular interest?

2. A right-angled triangle has the lengths of its sides in geometric progression. Find r, the common ratio of this geometric progression, correct to four decimal places.

3. Simplify
$$(1+x)(1+x^2)(1+x^4)(1+x^8)\ldots(1+x^{2^{n-1}}),$$
where x is any real number.

4. A segment of a circular disc has the dimensions shown. Find the area of the segment correct to four decimal places.

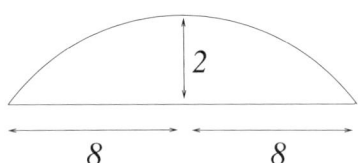

5. Evaluate
$$\sum_{r=1}^{n}[\tan r\theta \tan(r+1)\theta].$$

6. Write the sixth derivative of the function $\sec x$ as a polynomial in $s = \sec x$ only.

7. Prove that at least two different digits occur infinitely often in the decimal expansion of $\pi = 3.141592653589793\ldots$.

8. Evaluate
$$\int \frac{d\theta}{1+\sin\theta}.$$

9. Show how to cut a square into four rectangular pieces such that these four pieces can be reassembled to form two squares of different sizes.

10. Prove that the sum of the squares of five consecutive integers is not the square of an integer.

Twentieth Superbrain, 2003 — Problems

1. A mountaineer is one mile above sea-level. How far away is the furthest point the mountaineer can see on the horizon?

   (Assume that the Earth is a perfect sphere of radius 3960 miles.)

2. The lengths of the sides of a convex quadrilateral are 2, 3, 5 and 6 metres (as in the diagram) and the shorter diagonal has length four metres. Find, correct to four decimal places, the length of the longer diagonal.

   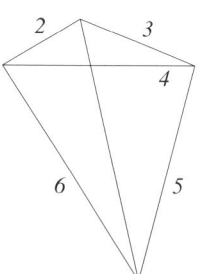

3. Find
$$\sum_{r=1}^{n} \frac{1}{\sin r\theta \sin(r+1)\theta}.$$

4. If p is a prime number, show that $2p^3 - 1$ and $2p^3 + 1$ cannot both be primes.

5. Every point in the plane is coloured red, blue or yellow. Show that there exist two points of the same colour which are a unit distance apart.

6. Find the value of $\lim_{n \to \infty} \dfrac{n}{2^n}$ and justify your answer.

7. Find all solutions in positive integers of the simultaneous equations $x + y = zw$ and $xy = z + w$.

8. How many rectangular 3×5 tiles can be fitted without overlap in a 17×31 rectangle?

9. The area of a triangle ABC is bisected by a line segment XY (with X on AB and Y on AC) whose length is as small as possible. Show that $|AX| = |AY|$.

10. The dimensions of an L-shaped figure are given. Locate the centre of gravity (centre of mass, centroid) of the figure. (Assume that the figure has uniform thickness and density.)

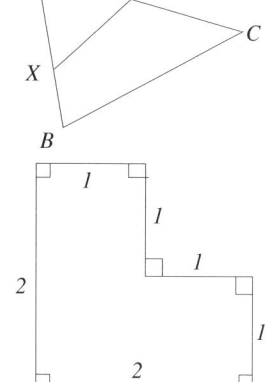

23

Twenty-First Superbrain, 2004 — Problems

1. In the Irish National Lottery, gamblers are asked to choose six different numbers from the set $\{1, 2, 3, \ldots, 42\}$. What percentage of choices contain at least two consecutive numbers?

2. Find all positive integers a and b such that
$$a^4 + (a+1)^4 + (a+2)^4 = b^4.$$

3. Prove or disprove that the map of the counties of Ireland (see opposite page) can be coloured with three different colours in such a way that counties which touch each other have different colours.

4. Evaluate $\displaystyle\sum_{n=1}^{\infty} \frac{\sin n\theta}{n!}$.

5. The area of an equilateral triangle OPQ is bisected by a curve AB of minimal length (where A is on OP and B on OQ). What is the equation of the curve with respect to the given axes?

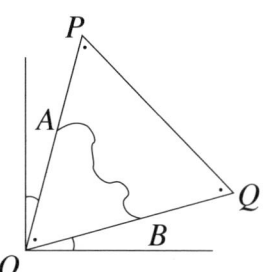

6. Evaluate $\displaystyle\int \frac{dx}{x^4 + 1}$.

7. Find the value of $\displaystyle\lim_{n \to \infty} \frac{2^n}{n!}$, and justify your answer.

8. Show how to cut this figure into three pieces and reassemble them to form a square.

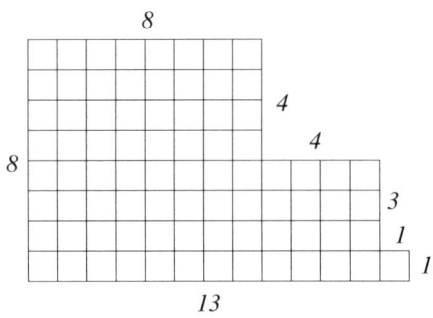

9. Solve the equation $x_1 + x_2 + \ldots + x_n = x_1 x_2 \ldots x_n$ for distinct natural numbers $x_1, x_2, \ldots, x_n$, with $n > 2$.

10. What are the coordinates of the point on the parabola $y^2 = 4x$ which is nearest to the point with coordinates $(-1, 4)$?

Twenty-Second Superbrain, 2005 — Problems

1. Four equal corners are cut from a square of side 10cm, to leave a regular octagon (i.e. an eight sided figure, all of whose sides have the same length and all of whose angles have the same measure). Find the area of the octagon, correct to three decimal places.

2. Prove that there are infinitely many squares ending in the digits 444.

3. Find the value of
$$\sum_{r=0}^{n} \cos(2r+1).$$

4. Evaluate
$$\int \sqrt{\tan x}\, dx.$$

5. Find each of the following:

   (a) $\lim\limits_{x \to \frac{\pi}{2}} [\sec x - \tan x]$. (b) $\lim\limits_{n \to \infty} \sqrt[n]{n}$.

6. If $AB \| XY \| DC$ and XY contains O, the intersection of AC and BD, find $|XY|$, given that $|AB| = 50$ and $|DC| = 70$.

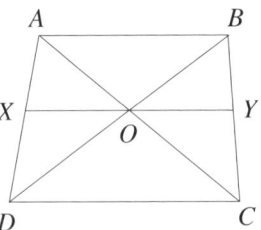

7. A, B and C are workers each of whom works at a steady but different rate. In fact A is the fastest, and C is the slowest. Any of the three working on her own takes a positive integer multiple of the time that the other two, working together, would take on the same job. If C is working on her own, how many times longer does she take on a job than A and B working together?

8. What is the minimum number of pieces into which this figure must be cut so that the pieces can be reassembled to form a square?

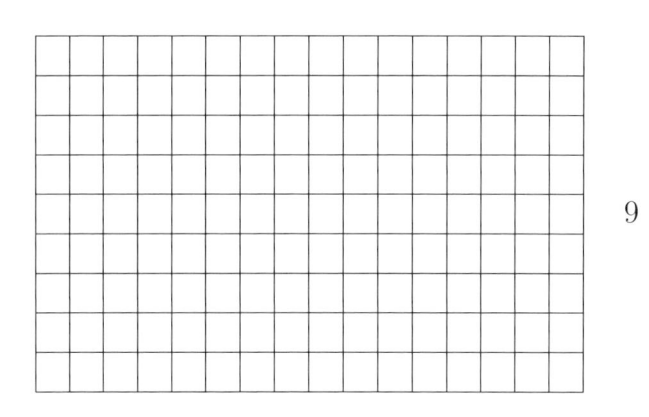

9. For all real numbers x, prove that
$$\sin(\cos x) < \cos(\sin x).$$

10. Evaluate $\displaystyle\prod_{n=2}^{\infty} \frac{n^3 - 1}{n^3 + 1} = \lim_{n \to \infty} \left[\frac{2^3 - 1}{2^3 + 1} \cdot \frac{3^3 - 1}{3^3 + 1} \cdot \frac{4^3 - 1}{4^3 + 1} \cdots \frac{n^3 - 1}{n^3 + 1} \right]$

Twenty-Third Superbrain, 2006 — Problems

1. Find all positive integers n such that $2^n + 1$ is the square of an integer.

2. A field is in the shape of a convex quadrilateral (i.e. no angle greater than 180°). The diagonals of the quadrilateral divide it into four triangles. Three of these triangles have areas 400 square metres, 700 square metres and 800 square metres. What is the largest area the field can have?

3. Evaluate the indefinite integral $\int \dfrac{dx}{x^n + x}$, where $n > 1$ is a positive integer.

4. Find all positive integers x and y with $x > y$ such that $\dfrac{1}{x} + \dfrac{1}{y} = \dfrac{1}{2006}$.

5. Evaluate
$$\sum_{r=1}^{n} \frac{1}{\sin(2^r x)},$$
where x is a real number and $2^n x$ is not a multiple of π.

6. Place the digits 1, 2, 3, 4, 5, 6, 7 and 8, one in each circle, in such a way that no two consecutive digits are placed in circles which are joined directly by a line segment.

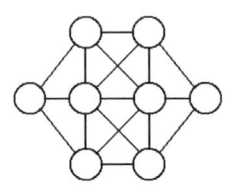

7. A water heating tank consists of an open hemisphere (radius r) surmounting an open top right-circular cylinder of radius r and height h. If the total surface area of the tank is a constant A, find the ratio $h : r$ which gives the maximum volume of the tank.

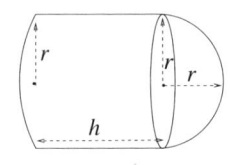

8. A_1, A_2, A_3, A_4 and A_5 are distinct points in the plane, each of whose coordinates is a pair of integers (i.e. lattice points). Show that there exists a point B and some line segment $A_i A_j$ such that B is a lattice point and B is an internal point of the line segment $A_i A_j$.

9. This figure has been cut into four congruent pieces as in the diagram. Show how to cut the figure into four congruent pieces of a different shape to those in the diagram. [Each angle is either $60°$ or $120°$.]

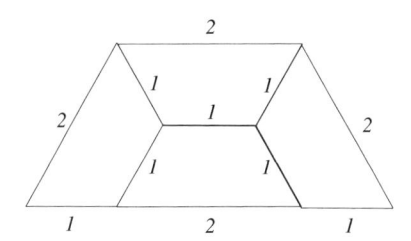

10. The triangular numbers $1, 3, 6, 10, 15, 21, \ldots$ are numbers of the form $\frac{n(n+1)}{2}$ for $n \in \mathbb{N}$. The square numbers $1, 4, 9, 16, 25, \ldots$ are numbers of the form n^2 for $n \in \mathbb{N}$.

Show that every triangular number greater than 1 is the sum of a square number and two triangular numbers.

Twenty-Fourth Superbrain, 2007 — Problems

1. Place the numbers 1, 2, 3, 4, 5, 6, 7 and 8 one at each corner of a cube in such a way that the sum of the numbers on each of the six faces of the cube is the same.

2. The positive integers $8 = 2^3$ and $9 = 3^2$ are consecutive integers each of which is a proper power (ie. of the form a^b where $a, b \in \mathbb{N}$ and $b \geq 2$). Show that there do not exist four consecutive integers each of which is a proper power.

3. You are given six rectangular tiles measuring $1 \times 2, 2 \times 3, 3 \times 4, 4 \times 5, 5 \times 6$ and 6×7 units. What is the area of the smallest integer-sided rectangle into which these tiles can be fitted without overlap? Justify your answer.

4. In the triangle ABC, X, Y, Z are points on the sides AC, AB and BC respectively such that AZ, BX and CY meet at an internal point O. Prove that
$$\frac{AX}{XC} + \frac{AY}{YB} = \frac{AO}{OZ}.$$

5. If $x + \frac{1}{x} = 2\cos\theta$, show that $x^n + \frac{1}{x^n} = 2\cos n\theta$, where n is an integer.

6. Find any set of positive integers $\{a, b, c, d\}$ such that $a^4 + b^5 + c^6 = d^7$.

7. Evaluate $\sum_{r=1}^{n} 2^r r! + \sum_{j=1}^{n+1} j 2^{j+1} j!$.

8. The volume of a closed right-circular cone of radius r and height h is 1000cm^3. If the radius is increasing at a rate of 3mm per second, at what rate is the total surface area A of the cone changing when $r = 5$cm?

9. Evaluate $\int_0^\pi \frac{x \sin x}{1 + \cos^2 x} dx.$

10. If a, b and c are the lengths of the sides of a triangle, find the least upper bound and greatest lower bound of
$$f(a, b, c) = \frac{a}{b+c} + \frac{b}{c+a} + \frac{c}{a+b},$$
and say if and when these values are attained.

30

Twenty-Fifth Superbrain, 2008 — Problems

1. If A, B and C are the angles of a triangle, prove that $\sin A + \sin B > \sin C$. Is this result true under the weaker assumption that $A+B+C = \pi$?

2. In how many different orders can ten people sit around a circular table if two particular people, Mister X and Miss Y, must never be seated next to each other?

3. If α, β and γ are the solutions of the equation $ax^3 + bx^2 + cx + d = 0$, and $a \neq 0$, find the value of $\alpha^3 + \beta^3 + \gamma^3$ in terms of a, b, c and d.

4. Evaluate $\int \left(\dfrac{\sin x}{x}\right)^2 dx - \int \dfrac{\sin 2x}{x} dx$.

5. Find all complex numbers z such that $(z+1)^7 = z^7 + 1$.

6. Evaluate $\lim\limits_{n\to\infty} \left[\dfrac{1}{n}\sqrt[n]{n!}\right]$.

7. If A and B are distinct points which lie inside a circular disc, show that every point on the line segment $[AB]$ also lies inside the circular disc.

8. Given that $\sum\limits_{n=1}^{\infty} \dfrac{1}{n^2} = \dfrac{\pi^2}{6}$, evaluate

$$\sum_{n=1}^{\infty} \dfrac{1}{n^2(n+1)^2}.$$

9. Factorise the expression $x^8 + 98x^4y^4 + y^8$ as a product of polynomials of degree at least one with integer coefficients which cannot be further factored in the same way.

10. Place the numbers 1, 2, 3, 4, 5 and 6 in the enclosed boxes in such a way that each digit appears once and only once in each row, each column, each main diagonal and in each of the six outlined sections.

Part 2:

Solutions

First Superbrain, 1984 — Solutions

1. Using each number exactly once, place the numbers 1, 2, 3, 4, 5, 6, 7, 8 and 9 in a 3 × 3 square so that all rows, columns and diagonals sum to different totals.

   Solution:

   Although this seems in many ways a less intuitive condition than the 'magic square' condition that all the sums be the same, it is actually far more common — 24960 (i.e. almost 7%) of the 362880 permutations of the first nine integers satisfy the condition, while only eight (corresponding to reflections and rotations of a unique solution) satisfy the magic square condition. As such, experimentation is probably the best way to solve this problem.

   It is possible to have as many as six numbers in the same position as they would be if the square was filled left to right and top to bottom. One such solution is:

   $$\begin{pmatrix} 1 & 2 & 3 \\ 4 & 5 & 9 \\ 6 & 8 & 7 \end{pmatrix}$$

   However, perhaps a more natural solution to find by experimentation is given by 'spiralling' out from the centre:

   $$\begin{pmatrix} 9 & 2 & 3 \\ 8 & 1 & 4 \\ 7 & 6 & 5 \end{pmatrix}$$

   Continuing the spiral also gives a solution to the same problem for a 5×5 square using the numbers from 1 to 25.

2. If x and y are positive integers, find all solutions of the equation $2xy - 4x^2 + 12x - 5y = 11$.

   Solution:

   Since the equation is linear in y, we can easily solve for y (note that the coefficient of y is never 0 for $x \in \mathbb{N}$):

   $$y = \frac{4x^2 - 12x + 11}{2x - 5} = 2x - 1 + \frac{6}{2x - 5}.$$

Now since $x, y \in \mathbb{N}$, it follows that $6/(2x - 5) = y - 2x + 1 \in \mathbb{N}$. For this condition to hold, we must have $2x - 5 \in \{\pm 1, \pm 3\}$ (since it is an odd divisor of 6), and hence $x \in \{1, 2, 3, 4\}$.

Only two of these values give rise to positive values of y, so the complete solution set is

$$(x, y) \in \{(3, 11), (4, 9)\}.$$

3. If ABC is an acute-angled triangle, show with proof how to find points S and R on the line BC, P on the line AB, and Q on the line AC, such that $PQRS$ is a square.

Solution:

The key to solving this problem is to weaken one of the restrictions on the square (in this case, that Q lies on AC) and find the locus of all possible points Q. This turns out to be simply a straight line, and so we can find the point which has the additional property of lying on AC by simply constructing this line and finding its intersection with AC.

We observe that

$$\tan \angle QBR = \frac{|QR|}{|BR|} = \frac{|QR|}{|SR| + |BS|} = \frac{1}{1 + \tan \angle ABC}.$$

So Q must lie on the intersection of a line through B with this slope (relative to BC) and the line AC. But the value of $\tan \angle QBR$ does not depend on the fact that Q is on AC — it holds for any square $PQRS$ with P on AB and R, S on BC.

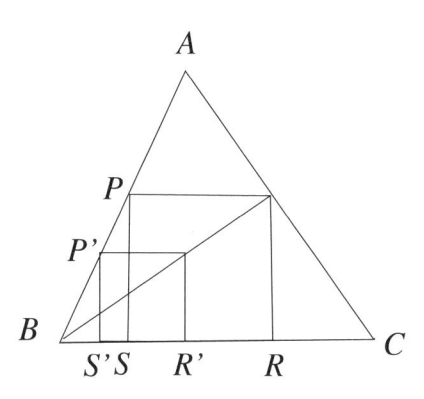

It is easy to construct such a square, say $P'Q'R'S'$. Pick P' arbitrarily on AB. Drop a perpendicular from P' to BC to get S'. Use compasses

35

(centre S', radius $|S'P'|$) to find R'. Construct lines parallel to $P'S'$ and $S'R'$ through R' and P' respectively — these intersect at Q'.

Now since BQ' and BQ have the same slope, Q must be simply the intersection of BQ' with AC.

Given Q, we can easily find P, R and S by the same method that P' was used to construct Q', R' and S': drop a perpendicular from Q to BC to get R, use a compasses (centre R, radius $|RQ|$) to find S, and draw a line through Q parallel to BC to get P (which is the intersection of this line with AB).

Alternative Solution:

The same method can obviously be used if we drop instead the assumption that P lies on AB. However, a slightly different solution in a similar vein can be achieved if we drop the assumption that S and R lie on BC, but instead simply require that SR be parallel to BC. Now, instead of constructing a square inside ABC, we construct one outside. In particular, we construct a square externally on BC, say $BCXY$.

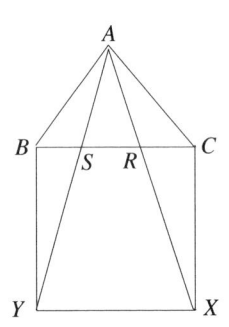

Then it is easy to confirm that the lines AY and AX are the loci of possible locations of S and R subject to the given assumptions. So the desired points S and R are simply the intersections of AY and AX respectively with BC, and the rest of the square may be easily constructed.

4. Ten books are arranged in a row on a shelf. How many ways can this be done if one particular book A must always be to the left, but not necessarily immediately to the left, of another book B?

Solution:

This can be solved by a simple counting argument, but it is much simpler to just take advantage of the symmetry of the situation: A will be to the left of B in exactly as many arrangements as B will be to the left of A, and hence the desired result is simply half the total number of ways to arrange the ten books — i.e. $\frac{10!}{2} = 1814400$.

Remark:

The key to this problem is to approach it from a point of view of probability rather than counting all possibilities — had the question been phrased in terms of probability, the answer would have been immediately obvious.

5. Assuming that
$$\lim_{x \to 0} \sqrt[x]{\frac{1+x}{1-x}}$$
exists, find its value.

Solution:

It is known that $\lim_{x \to 0} \sqrt[x]{1+x} = e$. But since x may approach zero from above or below, we may equally write $e = \lim_{x \to 0} \sqrt[-x]{1-x} = \lim_{x \to 0} \sqrt[x]{\frac{1}{1-x}}$.

Now if two functions both tend towards a finite limit, and the product of the two functions also tends toward a finite limit (all as $x \to 0$), then the limit of the product must be equal to the product of the limits.

So assuming that the given limit exists, we get

$$\lim_{x \to 0} \sqrt[x]{\frac{1+x}{1-x}} = \left(\lim_{x \to 0} \sqrt[x]{1+x} \right) \left(\lim_{x \to 0} \sqrt[x]{\frac{1}{1-x}} \right) = e^2.$$

Alternative Solution:

If we take the natural log of the given function, and then expand the log as a power series (for $0 < |x| < 1$), we get

$$\frac{1}{x} [\log(1+x) - \log(1-x)] = \frac{1}{x} \left[\sum_{n=1}^{\infty} \frac{(-1)^{n-1} x^n}{n} + \sum_{n=1}^{\infty} \frac{x^n}{n} \right]$$

$$= \frac{1}{x} \left[2 \sum_{n=0}^{\infty} \frac{x^{2n+1}}{2n+1} \right] = 2 \sum_{n=0}^{\infty} \frac{x^{2n}}{2n+1}.$$

Now letting x approach zero, it is easy to see that the log of the function approaches 2 (since all powers of x except x^0 vanish).

Finally, since log is continuous at 2, and since the limit of the function exists, the log of the limit must be equal to the limit of the log, and so

$$\log\left(\lim_{x\to 0} \sqrt[x]{\frac{1+x}{1-x}}\right) = \lim_{x\to 0} \log\left(\sqrt[x]{\frac{1+x}{1-x}}\right) = 2,$$

which gives the same result.

6. If $n = 2^k$ for $k \geq 1$, show that $\binom{n}{r}$, the number of combinations of n things r at a time, is an even number for $0 < r < n$.

Solution:

The nature of combinatorics problems is that there are typically many different ways to solve them, and that is certainly the case here. We offer two quite different solutions, but there are many other possible approaches.

One solution uses the binomial theorem and induction. The claim is equivalent to claiming that $(1+x)^{2^k}$ has all even coefficients except its first and last (which must both be 1). For $k = 1$, the claim is obviously true, so we proceed by induction.

Assume that the claim is true for $k = m$. But then,

$$(1+x)^{2^{m+1}} = (1+x)^{2^m}(1+x)^{2^m}.$$

Since each of the two terms on the right hand side has all coefficients even except their first and last, odd coefficients can arise only in their product when pairs of these are multiplied together. So, before grouping terms with the same power of x, there are four terms with odd coefficients:

$$1, \quad x^{2^m}, \quad x^{2^m}, \quad x^{2^{m+1}}.$$

But now, the middle two will group together, so that these can be ignored, and the first and last will be the first and last terms of the entire expression, so that the coefficients of all other terms must be even. So, if the claim is true for $k = m$, then it is true for $k = m+1$. By induction, the result follows.

Alternative Solution:

An even quicker solution can be achieved if the solver happens to be familiar with Legendre's formula: the exact power of a prime p dividing $n!$ is $(n - s_p(n))/(p-1)$, where $s_p(n)$ is the sum of the digits of n in base p. In particular, for $p = 2$, the exact power of two dividing $n!$

is simply $n - s_2(n)$, where $s_2(n)$ is the number of one's in the binary representation of n. It is a simple matter (a good exercise for a reader not already familiar with the result) to prove the validity of this formula by induction on n.

Applying this result, we get that the exact power of two dividing $\binom{n}{r}$ is $(n-1) - (r - s_2(r)) - (n - r - s_2(n-r)) = s_2(r) + s_2(n-r) - 1$ (since $n = 2^k$ has exactly one digit in base two).

But $s_2(m)$ is at least one for all positive integers m, so we must have $s_2(r) + s_2(n-r) - 1 \geq 1$, and so $2 \mid \binom{n}{r}$ for $0 < r < n$.

7. If $A = (a, b, c, d)$ is a set of four distinct elements, is it possible to define a closed binary operation $*$ on A such that the associative law $x * (y * z) = (x * y) * z$ never holds for any triple $x, y, z \in A$, equal or distinct?

Solution:

There are numerous solutions to this problem, but we present an elegant method, which is also quite simple to extend.

If we let $(a, b, c, d) = (x_0, x_1, x_2, x_3)$, and define $x_i * x_j = x_{i+1}$ (where the indices are taken mod 4), then we get $x_i * (x_j * x_k) = x_{i+1}$ and $(x_i * x_j) * x_k = x_{i+1} * x_k = x_{i+2}$. But x_{i+1} is not equal to x_{i+2} for any choice of i, j and k, and hence the required condition is met.

8. Let $\alpha = 1 + \frac{x^3}{3!} + \frac{x^6}{6!} + \ldots$, $\beta = x + \frac{x^4}{4!} + \frac{x^7}{7!} + \ldots$ and $\gamma = \frac{x^2}{2!} + \frac{x^5}{5!} + \frac{x^8}{8!} \ldots$ Assuming that all three series converge for $x \in \mathbb{R}$, prove that

$$\alpha^3 + \beta^3 + \gamma^3 = 1 + 3\alpha\beta\gamma.$$

Solution:

To solve this problem, we take advantage of a complex factorisation: $\alpha^3 + \beta^3 + \gamma^3 - 3\alpha\beta\gamma = (\alpha + \beta + \gamma)(\alpha + \omega\beta + \omega^2\gamma)(\alpha + \omega^2\beta + \omega\gamma)$, where ω is a complex cube root of unity.

But $\alpha + \beta + \gamma = 1 + x + \frac{x^2}{2} + \frac{x^3}{3!} + \ldots = \sum_{n=0}^{\infty} \frac{x^n}{n!} = e^x$,
$\alpha + \omega\beta + \omega^2\gamma = 1 + \omega x + \frac{\omega^2 x^2}{2} + \frac{x^3}{3!} + \ldots = \sum_{n=0}^{\infty} \frac{(\omega x)^n}{n!} = e^{\omega x}$,
and $\alpha + \omega^2\beta + \omega\gamma = 1 + \omega^2 x + \frac{\omega x^2}{2} + \frac{x^3}{3!} + \ldots = \sum_{n=0}^{\infty} \frac{(\omega^2 x)^n}{n!} = e^{\omega^2 x}$.

Combining these, we get $\alpha^3 + \beta^3 + \gamma^3 - 3\alpha\beta\gamma = e^x e^{\omega x} e^{\omega^2 x} = e^{x(1+\omega+\omega^2)}$. Now, $\omega^3 - 1 = (\omega - 1)(\omega^2 + \omega + 1) = 0$, and $\omega \neq 1$, so $\omega^2 + \omega + 1 = 0$, and hence $e^{x(1+\omega+\omega^2)} = e^0 = 1$, so $\alpha^3 + \beta^3 + \gamma^3 = 3\alpha\beta\gamma + 1$ as required.

9. Prove that $\cos 29°$ is not a rational number.

   Solution:

   One way of proving that a number is irrational is to find a polynomial with integer coefficients of which it is a root, and apply the rational root theorem to generate a finite set of possible rational values, which can then be checked. So we seek a polynomial with integer coefficients of which $\cos 29°$ is a root.

   Such a polynomial can be found using de Moivre's Theorem:
   $$\begin{aligned}\cos(90\theta) &= \operatorname{Re}\left[(\cos\theta + i\sin\theta)^{90}\right] \\ &= \sum_{n=0}^{45}(-1)^n \cos^{90-2n}\theta \sin^{2n}\theta \\ &= \sum_{n=0}^{45}(-1)^n \cos^{90-2n}\theta \left(1 - \cos^2\theta\right)^n.\end{aligned}$$

   But $\cos(90 \times 29°) = \cos(7 \times 360° + 90°) = 0$. So Substituting $\theta = 29°$, we get that $c = \cos 29°$ must be a root of the polynomial
   $$\sum_{n=0}^{45}(-1)^n c^{90-2n}\left(1 - c^2\right)^n,$$
   which has integer coefficients.

   Although the expression looks intimidating, we only need a small amount of information to apply the rational root theorem. Letting $c = 0$ gives us the constant term as $(-1)^{45} = -1$, and so any rational root $\frac{a}{b}$ of the polynomial (with a and b coprime) must have $a | -1$, and so be of the form $\pm\frac{1}{m}$ for some $m \in \mathbb{N}$. But, since cos is decreasing on $\left(0, \frac{\pi}{2}\right)$, $\frac{1}{1} = \cos 0° > \cos 29° > \cos 30° = \frac{1}{2}$, and so $\cos 29°$ is not of the given form, and hence is not a rational number.

10. Form a nine digit number using each of the digits 1, 2, 3, 4, 5, 6, 7, 8, 9, once and only once so that the number formed by the nine digits is a multiple of nine, the number formed by the first eight digits is a multiple of eight, the number formed by the first seven is a multiple of seven and so on, i.e. the number formed by the first n digits is a multiple of n, for $1 \leq n \leq 9$.

    Solution:

    The best approach to a problem like this is a healthy mix of abstract simplifications and experimentation.

A few simple observations can speed things up. Firstly, to be a multiple of two, four, six or eight, a number must end in an even digit, and hence the four even digits are in the four even-indexed positions. To be a multiple of five, a number must end in five, so the fifth digit is five. For the first three digits to form a multiple of three, their sum must be a multiple of three, and similarly the sum of the next three digits, and of the last three digits is a multiple of three.

Taking the number formed by the first four digits mod 4, we see that $1000a_1 + 100a_2 + 10a_3 + a_4 \equiv 2a_3 + a_4 \equiv 0$, where a_i is the ith digit. So, since a_3 is odd, we must have $a_4 \equiv 2 \bmod 4$, so $a_4 = 2$ or 6. Similarly, $2a_7 + a_8 \equiv 0 \bmod 4$, and a_7 is also odd, so we must have $\{a_4, a_8\} = \{2, 6\}$. Now since $a_4 + a_5 + a_6 = 7 + a_6$ or $11 + a_6$ must be a multiple of three, and a_6 must be even, we see that $a_4 = 2 \Rightarrow a_6 = 8$, $a_2 = 4$, and $a_4 = 6 \Rightarrow a_6 = 4$, $a_2 = 8$.

So the two possible forms of the number are *4*258*6* and *8*654*2*. In the first case, the divisibility of the last triple by three requires that it be 369 or 963, and the divisibility by eight requires that it be 963, but neither of the two numbers of this form are solutions. In the second case, to satisfy the divisibility by three, the first two blanks must be filled by one of $\{1, 7\}$ ($\equiv 1 \bmod 3$) and one of $\{3, 9\}$ ($\equiv 0 \bmod 3$), in either order. This gives a total of sixteen different numbers. Trial and error shows that exactly two of these numbers satisfy the condition of divisibility by seven — namely 381654729 and 783654921. Finally, the second of these two does not satisfy the condition of divisibility by eight, and so 381654729 is the unique solution.

Second Superbrain, 1985 — Solutions

1. In a darts competition, each dart scores 40, 39, 24, 23, 17 or 16 points. How many darts must be thrown to get exactly 100 points?

   Solution:

   Note that all of the darts are either a multiple of eight, or one above or below a multiple of eight. This suggests that considering the remainders on division by eight (which will all be small) may be a good way to reduce the problem.

   Taking the values of each of the darts modulo eight, we get $\{16, 24, 40\} \equiv \{0\}$, $\{23, 39\} \equiv \{-1\}$ and $\{17\} \equiv \{1\}$. Suppose that a darts get scores in the first set, b darts get scores in the second set, and c darts get scores in the third set. Then $c - b \equiv 100 \equiv 4 \bmod 8$. So at least one of b and c must be at least four.

   But if $b \geq 4$, then the sum of the darts scoring in the second set is either 92 or more, and it is then impossible to score 100. So $c \geq 4$, and so at least four darts score 17. The total of these four is 68, which means that some number of other darts must score the remaining 32. But now, no one dart can score 32, and two darts will score at least 32, with equality if and only if they are both 16 (i.e. the minimum score).

   So, the only way to score 100 is with four 17s and two 16s, for a total of six darts.

2. Five points lie inside an equilateral triangle of side 2 units. Prove that at least two of the points are no more than a unit distance apart.

   Solution:

   The key to solving this problem is to divide up the triangle into four smaller congruent triangles (this can be done by connecting the midpoints of each of the sides):

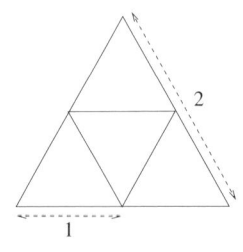

   Now, there are five points contained in four triangles. By the pigeonhole principle, at least one of the triangles must contain at least two points.

But any two points in an equilateral triangle are separated by a distance of at most the sidelength of the triangle, so the two points must be at most unit distance apart, as required (since each of the four sub-triangles has a sidelength of one unit).

3. Find all prime numbers that can be written in the form $a^4 + 4b^4$, where a and b are positive integers.

Solution:

Since we are asked to find all solutions, the natural guess should be that there are only finitely many — and so our task is to prove that the given expression is in general *not* prime, and to find the exceptions to the rule. The easiest way to do this is to seek a factorisation. We can do this by 'completing the square':

$$\begin{aligned} a^4 + 4b^4 &= (a^4 + 4a^2b^2 + 4b^4) - (4a^2b^2) \\ &= (a^2 + 2b^2)^2 - (2ab)^2 \\ &= (a^2 - 2ab + 2b^2)(a^2 + 2ab + b^2). \end{aligned}$$

But now, $a^2 + 2ab + 2b^2 > a^2 - 2ab + 2b^2 = (a-b)^2 + b^2 \geq 1$, with equality on the left if and only if $a = b = 1$. But since $a^4 + 4b^4$ has been factorised, it can be prime only if its smaller divisor is one — so we must have $a^2 - 2ab + 2b^2 = 1$, and hence we must have $a = b = 1$.

So the only prime expressible in the given form is $5 = 1^4 + 4.1^4$.

4. If A, B and C are angles with

$$\sin A + \sin B + \sin C = 0 = \cos A + \cos B + \cos C,$$

prove that

$$\cos 3A + \cos 3B + \cos 3C = 3\cos(A + B + C).$$

Solution:

On looking at this problem, it may be tempting to take a brute force approach, by expanding each of the cosine terms and simplifying, but this is a lengthy and very messy method. A more useful insight may be had by considering the similarity between the given equation and the well-known fact $a + b + c = 0 \Rightarrow a^3 + b^3 + c^3 = 3abc$ (this can easily be confirmed by factoring).

In order to make the first equation look more like the second, we use the identity $\cos x = \operatorname{Re}(e^{ix})$ (where $\operatorname{Re}(z)$ is the real part of z) to re-write it as $\operatorname{Re}\left(e^{3iA} + e^{3iB} + e^{3iC}\right) = 3\operatorname{Re}\left(e^{i(A+B+C)}\right)$. Now this equation will certainly be true if $e^{3iA} + e^{3iB} + e^{3iC} = 3e^{i(A+B+C)}$, and with the above condition it is sufficient to show that $e^{iA} + e^{iB} + e^{iC} = 0$. But $e^{iA} + e^{iB} + e^{iC} = (\cos A + \cos B + \cos C) + i(\sin A + \sin B + \sin C) = 0$, so the result is proved.

Alternative Solution:

The solution can be reformulated using de Moivre's Theorem to avoid the exponential. To use $a + b + c = 0 \Rightarrow a^3 + b^3 + c^3 = 3abc$, we set $a = \cos A + i \sin A$, $b = \cos B + i \sin B$, $c = \cos C + i \sin C$. Then $a + b + c = 0$ follows from the given conditions, and so we have

$$a^3 + b^3 + c^3 = \cos 3A + \cos 3B + \cos 3C + i(\sin 3A + \sin 3B + \sin 3C)$$
$$= 3abc = 3\cos(A+B+C) + 3i\sin(A+B+C).$$

Separating real and imaginary parts gives the desired result, and also

$$\sin 3A + \sin 3B + \sin 3C = 3\sin(A + B + C).$$

5. In how many different ways is it possible to pay a punt (100p) using 50p, 10p and 5p pieces only?

Solution:

This problem is very easily solved with a 'top-down' approach. Consider first the number of 50p pieces, then the number of 10p pieces.

Obviously we cannot pay more than two 50p pieces, and if we pay two then there is no more to pay, and so there is exactly one way to pay with two 50p pieces.

If we pay exactly one 50p piece, then we also pay between zero and five 10p pieces, and each of these defines the number of 5p pieces to be paid to make up 100p, so there are exactly six ways to pay with one 50p piece. Similarly, there are exactly eleven ways to pay with no 50p pieces (and hence between zero and ten 10p pieces).

So in total, there are eighteen ways to pay 100p with 50p, 10p and 5p pieces, corresponding to the following integer triples ({50p,10p,5p}):

{{2,0,0}, {1,5,0}, {1,4,2}, {1,3,4}, {1,2,6}, {1,1,8}, {1,0,10}, {0,10,0}, {0,9,2}, {0,8,4}, {0,7,6}, {0,6,8}, {0,5,10}, {0,4,12}, {0,3,14}, {0,2,16}, {0,1,18}, {0,0,20}}

6. If, in the accompanying figure, $ABDC$, $BEFD$, $EGHF$ are squares, find, with proof, the size of angle $\alpha + \beta$.

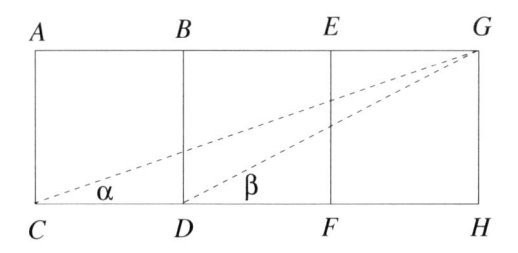

Solution 1:

A trigonometric approach to this problem yields a quick answer:

$$\alpha + \beta = \tan^{-1}(\tan(\alpha + \beta)) = \tan^{-1}\left(\frac{\tan\alpha + \tan\beta}{1 - \tan\alpha\tan\beta}\right)$$

$$= \tan^{-1}\left(\frac{\frac{1}{2} + \frac{1}{3}}{1 - \left(\frac{1}{2}\right)\left(\frac{1}{3}\right)}\right) = \tan^{-1}(1) = \frac{\pi}{4}.$$

Solution 2:

However, a geometric approach also yields a nice solution:

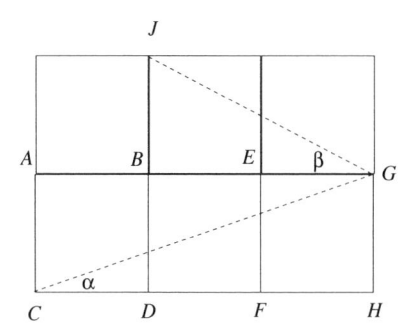

Since all the figures are squares, we can see that $|GJ| = \sqrt{5}|AB|$, $|CJ| = \sqrt{5}|AB| = |GJ|$ and $|CG| = \sqrt{10}|AB| = \sqrt{2}|GJ|$. This means that the the triangle CGJ has sidelengths proportional to $(1, 1, \sqrt{2})$, and so $\angle JCG = \angle CGJ = \frac{1}{2}\angle CJG = \frac{\pi}{4}$. But $\alpha + \beta = \angle CGJ$, so $\alpha + \beta = \frac{\pi}{4}$.

7. Given a triangle ABC, show how to find a point P such that
$$|PA|^2 + |PB|^2 + |PC|^2$$
is as small as possible.

Solution:

Let $A = (x_1, y_1)$, $B = (x_2, y_2)$, $C = (x_3, y_3)$, $P = (x, y)$ and
$$\Sigma = |PA|^2 + |PB|^2 + |PC|^2 = \sum_{n=1}^{3}\left[(x - x_n)^2 + (y - y_n)^2\right].$$

Completing the square in x and y, we get
$$\begin{aligned}\Sigma &= 3\left(x - \tfrac{1}{3}\sum_n x_n\right)^2 + \sum_n x_n^2 - \tfrac{1}{3}\left(\sum_n x_n\right)^2 \\ &\quad + 3\left(y - \tfrac{1}{3}\sum_n y_n\right)^2 + \sum_n y_n^2 - \tfrac{1}{3}\left(\sum_n y_n\right)^2 \\ &= 3\left(x - \tfrac{1}{3}\sum_n x_n\right)^2 + 3\left(y - \tfrac{1}{3}\sum_n y_n\right)^2 \\ &\quad + \sum_n \left(x_n - \tfrac{1}{3}\sum_i x_i\right)^2 + \sum_n \left(y_n - \tfrac{1}{3}\sum_i y_i\right)^2 \\ &\geq \sum_n \left(x_n - \tfrac{1}{3}\sum_i x_i\right)^2 + \sum_n \left(y_n - \tfrac{1}{3}\sum_i y_i\right)^2,\end{aligned}$$
with equality if and only if $x = (x_1 + x_2 + x_3)/3$ and $y = (y_1 + y_2 + y_3)/3$. But now, the point whose co-ordinates are the mean of the co-ordinates of the vertices of the triangle is the centroid of the triangle. So, the point P which minimizes the given sum must be the centroid.

We can construct P by joining two vertices to the centres of the opposite sides; P is the intersection of the two lines produced.

Alternative Solution:

If the reader is familiar with partial differentiation, it is possible to analyse the minima of the function Σ defined in the previous solution by that method. Moreover, note that Σ clearly cannot be minimal for a point P outside the triangle, since all three of $|PA|, |PB|$ and $|PC|$ could be reduced by bringing P closer to the triangle. So, to find the global maximum of Σ, it is sufficient to find the maximum of Σ on the (compact) set of points within the triangle.

Fixing y and differentiating Σ with respect to x gives
$$\frac{\partial \Sigma}{\partial x} = 2\sum_{n=1}^{3}(x - x_n) = 6\left(x - \frac{x_1 + x_2 + x_3}{3}\right).$$

But $\dfrac{\partial^2 \Sigma}{\partial x^2} = 6 > 0$, so $x = \dfrac{x_1 + x_2 + x_3}{3}$ gives a minimum value for Σ. Similarly, $y = \dfrac{y_1 + y_2 + y_3}{3}$ gives the minimum with respect to y. So, to minimise Σ, P must be the centroid of the triangle ABC, as above.

8. What is the maximum and the minimum number of Friday the thirteenths that can occur in any calendar year?

   Solution:

   If we name the days of the week 1, 2, 3, 4, 5, 6 and 7 in order, with day 1 being the day of January the thirteenth, then we can determine the day of the thirteenth of every month (for both a regular and a leap year):

	Day 1	Day 2	Day 3	Day 4	Day 5	Day 6	Day 7
Regular:	Jan Oct	May	Aug	Feb Mar Nov	Jun	Sept Dec	Apr Jul
Leap:	Jan Apr Jul	Oct	May	Feb Aug	Mar Nov	Jun	Sept Dec

   So, over the entire year (in either case) any day occurs between one and three times on the thirteenth of a month, and so in particular, between one and three Fridays fall on the thirteenth in any year.

   However, it is an easy exercise to confirm that any twenty-eight year period (not including the turn of a century, since this changes the pattern) will include examples of all fourteen combinations of possible days for the thirteenth of January and regular and leap years, so the upper and lower limits are in fact the maximum and minimum.

   Remark:

   The reader with some knowledge of abstract algebra should immediately recognize the method of splitting years into equivalence classes, where two years are considered equivalent if and only if each date in each year falls on the same day of the week.

9. By considering
$$\int_0^1 \frac{x^4(1-x)^4}{1+x^2}\,dx,$$
show that $\pi < \frac{22}{7}$.

   Solution:

   To simplify the integration, we use long division:
   $$\frac{x^4(1-x)^4}{1+x^2} = x^6 - 4x^5 + 5x^4 - 4x^2 + 4 - \frac{4}{x^2+1}.$$

Substituting this and integrating, we get

$$\int_0^1 \frac{x^4(1-x)^4}{1+x^2} = \frac{1}{7} - \frac{4}{6} + \frac{5}{5} - \frac{4}{3} + 4 - 4\left(\frac{\pi}{4}\right) = \frac{22}{7} - \pi.$$

But the integrand is positive for all $x \in (0,1)$, and so the integral must be positive, which gives $\frac{22}{7} > \pi$.

Remark:

We can get a corresponding lower bound on π by noting that $x(1-x) \leq \frac{1}{4}$ for $0 \leq x \leq 1$:

$$\frac{22}{7} - \pi = \int_0^1 \frac{x^4(1-x)^4}{1+x^2} dx < \frac{1}{4^4} \int_0^1 \frac{dx}{1+x^2} = \frac{\pi}{4^5}.$$

So, we have

$$0.999 < \frac{1}{1+\frac{1}{4^5}} < \frac{7\pi}{22} < 1.$$

10. In this long division each asterisk stands for a whole number. Reconstruct all the calculations, given that there is no remainder.

```
              * 7 * * *
    * * * |* * * * * * *
           * * * *
             * * *
             * * *
             * * * *
             * * *
                * * * *
                * * * *
```

Solution:

Clearly the divisor is a three-digit number, but seven times it is also a three-digit number, so its first digit must be 1. Using this, we can make a few simple substitutions:

```
                    a 7 b 0 c
    1 x y | * * * * * * * *
            * * * *
              * * *
              * * *
              1 0 d *
                9 e *
                  1 * * *
                  1 * * *
```

Now $a \times 1xy$ and $c \times 1xy$ are four-digit numbers, while $7 \times 1xy$ is a three digit number, so a and c must be 8 or 9. $10d* - 9e* \leq 19$, so d must be 0 or 1 and e must be 8 or 9. Finally, $b \times 1xy$ starts with a 9, so b must be either 7 or 8.

Now, $980 < 9e* < 1000$, and $9e*$ is a multiple of either 7 or 8. The only such numbers in the range are 987, 984, 992 and 994, corresponding to values of $1xy$ of 123, 124, 141 or 142. However, we note that if $7 \times 1xy$ began with a 9, then the remainder when it was subtracted from a three-digit number could not be another three-digit number, which rules out 141 and 142.

So $1xy$ must be 123 or 124, and since $980 < b \times 1xy < 1000$, we must have $b = 8$. Also, since $a \times 1xy$ and $c \times 1xy$ are both four-digit numbers, we must have $a = c = 9$.

So we have the exact value of $a7b0c$, which must be 97809 and two possible values for $1xy$, which must be 123 or 124. It is easy to check the two possibilities to confirm that only one yields a valid solution:

```
              9 7 8 0 9
    1 2 4 | 1 2 1 2 8 3 1 6
            1 1 1 6
              9 6 8
              8 6 8
              1 0 0 3
                9 9 2
                  1 1 1 6
                  1 1 1 6
```

49

Third Superbrain, 1986 — Solutions

1. 1986 is the year Cork 801. Show how to write the number 801 using each of the digits 0, 1, 2, 3, 4, 5, 6, 7, 8, 9 once and only once, using any mathematical operation you wish.

   Solution:

   There are a huge selection of solutions to this problem — and it is made especially easy by the inclusion of the digit 0, which means that we can effectively remove the requirement of using each digit once (by multiplying any left-over digits by 0).

   A natural approach is to take advantage of the fact that 801 is very close to 800, which factors easily. For instance, we might write $801 = 1 + (8.4.5.(3+2)) + 0.6.7.9$.

   The problem is slightly harder if we omit the zero, but there are still many solutions. For instance $801 = 9.(8.(7+4)+1) + (2+6) - (5+3)$.

2. What, in your opinion is the next term of the sequence, and why?

   $$F, 4, E, S, 9, S, E, 5, E, \ldots$$

   Solution:

   The key to solving this problem is to reconcile the fact that the sequence is made up of both numbers and letters. One approach which the solver might consider would be to convert the numbers into letters by writing them as Roman numerals (I, II, III, IV, V, VI, VII, VIII, IX, X, etc.):

   $$F, IV, E, S, IX, S, E, V, E$$

   Now, with a regrouping of the terms, the pattern, and also the next term should be clear:

   $$FIVE, SIX, SEVE\ldots$$

   The sequence is simply spelling out the numbers from five up, so the missing next term is simply the last letter of seven, i.e. N.

3. Find the value of $\cos^2 0° + \cos^2 1° + \ldots + \cos^2 89°$ correct to twenty decimal places.

Solution:

The key to this problem, as with a great many problems involving cosine, is to take advantage of the close relationship between sine and cosine — in particular, the symmetry of the two over the range $(0°, 90°)$.
Let $S = \sum_{i=0}^{89} \cos^2 i°$. Since $\cos x = \sin(90° - x)$, we have $S = \sum_{i=1}^{90} \sin^2 i°$. But $\cos 90° = \sin 0° = 0$, so we have

$$S = \sum_{i=0}^{90} \cos^2 i° = \sum_{i=0}^{90} \sin^2 i° = \frac{1}{2} \sum_{i=0}^{90} \left(\cos^2 i° + \sin^2 i°\right) = \frac{\sum_{i=0}^{90} 1}{2} = 45.5.$$

4. Which of the following numbers is bigger, and why?

$$\int_0^1 \sqrt[4]{1 - x^7}\,dx \text{ or } \int_0^1 \sqrt[7]{1 - x^4}\,dx$$

Solution:

The integral on the left is the area under the curve $y = \sqrt[4]{1 - x^7}$ on the range $[0, 1]$. The equation for the curve can be rewritten as $y^4 + x^7 = 1$, in which case we see that the integral is simply the area enclosed by this curve and the positive x and y axes.

Similarly, the integral on the right is the area enclosed by the curve $y^7 + x^4 = 1$ and the positive x and y axes.

But now, these two curves are identical up to a swapping of variables — or, geometrically speaking, up to a reflection in the line $x = y$. But since area is preserved under reflection, the two areas, and hence the two integrals, must be equal.

Remark:

This solution clearly has more general application. In particular, it can be used to generalise the result to

$$\int_0^1 \sqrt[p]{1 - x^q}\,dx = \int_0^1 \sqrt[q]{1 - x^p}\,dx$$

for any positive integers p and q.

In fact it can be applied to any curve which is a function of both x and y on $[0, 1]$ and which goes from $(x, y) = (0, 1)$ to $(x, y) = (1, 0)$. We also present an alternative, non-geometric proof of the general case (which can, in particular, be applied to the given problem).

Alternative Solution:

Let f and g be functions defined on $[0, 1]$ such that

(a) $f(0) = g(0) = 1$,

(b) $f(1) = g(1) = 0$,

(c) $f(g(x)) = x = g(f(x))$ for all $x \in [0,1]$.

Let $u \in (0,1)$ and let $x = g(u)$ (so $f(x) = f(g(u)) = u$). Then by a substitution, we get

$$\int_0^1 f(x)dx = \int_{f(0)}^{f(1)} ug'(u)du = -\int_0^1 ug'(u)du.$$

Integrating by parts, we get

$$\int_0^1 f(x)dx = -\int_0^1 ug'(u)du = -[ug(u)]_0^1 + \int_0^1 g(u)du = \int_0^1 g(u)du.$$

Setting $f(x) = \sqrt[4]{1-x^7}$ and $g(x) = \sqrt[7]{1-x^4}$, we get the same particular result back.

5. Find the value of
$$\sum_{n=1}^{\infty} \frac{2n+1}{n^4 + 2n^3 + n^2}.$$

Solution:

To an experienced mathematician, the coefficients of the denominator of the fraction in the sum (1, 2, 1 on consecutive decreasing powers) should be a strong reminder of the identity $(x+1)^2 = x^2 + 2x + 1$ — although even if the link is not immediately apparent, the denominator is easily factored as $n^4 + 2n^3 + n^2 = n^2(n+1)^2$.

But as soon as we think of the denominator as being the product of n^2 and $(n+1)^2$, we should also recognise that $2n+1$ is precisely the difference of these two. This means that the fraction is of the form $\frac{a-b}{ab} = \frac{1}{b} - \frac{1}{a}$, with $a = (n+1)^2$ and $b = n^2$.

But now, we can write

$$\frac{2n+1}{n^4 + 2n^3 + n^2} = f(n) - f(n+1),$$

with $f(n) = 1/n^2$, giving us a telescoping sum. So

$$\sum_{n=1}^{k} \frac{2n+1}{n^4 + 2n^3 + n^2} = f(1) - f(k+1) = 1 - \frac{1}{(k+1)^2},$$

and hence
$$\sum_{n=1}^{\infty} \frac{2n+1}{n^4 + 2n^3 + n^2} = \lim_{k \to \infty} \left(1 - \frac{1}{k^2}\right) = 1.$$

6. A thin rod is broken into three pieces. What is the probability that a triangle can be formed from the three pieces?

Solution:

Suppose that the rod has a length of one, and that the three pieces have lengths a, b and $1 - a - b$. Then the space of possible values of (a, b) in the $a - b$ plane is the triangle formed by the intersection of the half planes $a > 0$, $b > 0$ and $a + b < 1$. If we assume that each pair in the space is equally likely (i.e. that the stick is broken arbitrarily) then the probability that a triangle can be formed is simply the ratio of the area of the set of pairs (a, b) for which a triangle can be formed to the area of the entire triangle (which is simply $1/2$).

Now, a necessary and sufficient condition for the three pieces to form a triangle is that the three sides satisfy the triangle inequalities — that is, that the sum of the lengths of each pair be longer than the third. Equivalently, each of a, b and $1 - a - b$ must be less than $1/2$ (i.e. half the perimeter).

The area in which all three inequalities hold is the triangle formed by the intersection of the three half planes $a < 1/2$, $b < 1/2$ and $a + b > 1/2$, which has area $1/8$.

Taking the ratio of the two, we get that the probability of a triangle being formed is $2/8 = 1/4$.

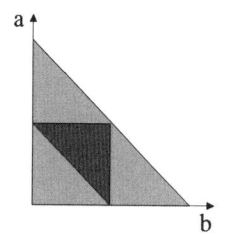

7. Prove that $\sum_{r=1}^{n} \frac{1}{r}$ is not an integer for $n > 1$.

Solution:

We know that, for each positive integer r less than or equal to n, we can write $1/r = k_r/n^*$, where n^* is the least common multiple of the first n integers and k_r is an integer.

But for all integers $n > 1$, there exists a (unique) positive integer k such that $2^k \leq n < 2^{k+1}$. This means that k must be the exact power of two dividing n^*. We write $n^* = 2^k b$, where b is an odd integer. Moreover, since the only number less than or equal to n which is divisible by 2^k is 2^k, k_r must be even for all r not greater than n except 2^k, and odd for $r = 2^k$.

Now, if we write $\sum_{r=1}^{n} \frac{1}{r} = \frac{\sum_{r=1}^{n} k_r}{2^k b}$, we can see that the numerator is odd (since it is a sum of integers, exactly one of which is odd), while the denominator is even, and hence the entire sum cannot be an integer.

8. A right-angled triangle has all of its sides an integer length. If the length of the perimeter equals the area, find all such triangles.

Solution:

Let the two perpendicular sides have lengths a and b. Then the area is $ab/2$, and the perimeter is $a + b + \sqrt{a^2 + b^2}$, and so the condition becomes $ab - 2a - 2b = 2\sqrt{a^2 + b^2}$.

Squaring and cancelling common terms, we get $a^2b^2 - 4a^3b - 4ab^3 + 8ab = ab(ab - 4a - 4b + 8) = 0$. But a and b must be positive, so we have $ab - 4a - 4b + 8 = 0$, which we can partially factor to get $(a-4)(b-4) = 8$.

Now since 8 can be written as only two distinct integer products, and since the factorisation is symmetric in the two variables, we must have $\{a - 4, b - 4\} = \{1, 8\}$ or $\{2, 4\}$, giving exactly two possible triangles:

$$(5, 12, 13) \text{ and } (6, 8, 10).$$

It is easy to confirm that both of these triangles do indeed have the desired property.

Alternative Solution:

We can take a slightly different approach by using the well-known formula for integer Pythagorean triples,

$$(a, b, c) = \left(2mnr, r(m^2 - n^2), r(m^2 + n^2)\right),$$

where a and b are the perpendicular sides, and c is the hypotenuse, and where m, n and r are positive integers, with $m > n$, and where m and n are coprime and have different parity (i.e. one is odd and the other is even).

Substituting these into the formulae for area $(ab/2)$ and perimeter $(a + b + c)$, the condition becomes

$$mn(m^2 - n^2)r^2 = 2mr(m + n).$$

Cancelling common factors on both sides, we get $nr(m - n) = 2$. Since $m - n$ is odd, we must have $m - n = 1$, and hence $\{n, m, r\} = \{2, 3, 1\}$ or $\{1, 2, 2\}$, which gives the same two solutions as before.

9. Is it possible to have a right-circular cone in which the total surface area is equal to the volume? Justify your answer.

   Solution:

   If the base of the cone has radius r and the apex is a height h above the base, then we know that the volume of the cone is $V = \frac{1}{3}\pi r^2 h$, while its surface area is $S = \pi r(r + \sqrt{r^2 + h^2})$. If we suppose that the radius and height are equal, we get

   $$\frac{V}{S} = \frac{r}{3(1 + \sqrt{2})},$$

   so that $r = 3(1 + \sqrt{2})$ gives $V = S$, and hence the result is proved.

   Although the assumption $r = h$ simplifies the calculations, it is not a necessary condition. In fact, it is easy to confirm that in general, if the ratio $\frac{h}{r}$ has the value k, then

   $$r = \frac{3}{k}\left(1 + \sqrt{1 + k^2}\right)$$

   gives $V = S$, and hence such cones exist for every possible ratio of radius to height.

10. Find the smallest positive integer that can be expressed as the sum of two squares of positive integers in two different ways.

    Solution:

    The easiest way to approach this question is to experiment — construct a table indicating all possible sums $a^2 + b^2$ with a and b less than n, and increase n until a repeat appears.

	1	2	3	4	5	6	7	8
1	2	5	10	17	26	37	50	65
2		8	13	20	29	40	53	68
3			18	25	34	45	58	73
4				32	41	52	65	80
5					50	61	74	89
6						72	85	100
7							98	113
8								128

    So we see that the smallest solution is $5^2 + 5^2 = 7^2 + 1^2$, while the smallest solution without repetition is $7^2 + 4^2 = 8^2 + 1^2$.

Fourth Superbrain, 1987 — Solutions

1. What, in your opinion, is the next term in the sequence

$$3, 5, 1, 15, 11, 10$$

   and why?

   Solution:

   Although it is very hard to approach a question like this in an entirely systematic manner, a first step might be taken by noticing that the first and second halves of the sequence are each 'almost' a decreasing triple of integers (3, 2, 1 and 12, 11, 10). We might proceed by pursuing this similarity.

   The second thing we might notice is that the only change required to restore the two triples is to replace the 5s in each of them with a 2. In order to increase this apparent similarity, we might think of the digital representations of the two numbers:

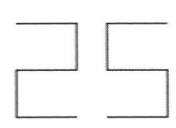

   We note that either a vertical or a horizontal reflection of one gives rise to the other. However, if we now look at the remaining digits in the sequence, we note that each of their digital representations is invariant under a reflection in their horizontal axes. So, if we write the entire sequence in digital form and then reflect it in a horizontal axis, we get the new sequence 3, 2, 1, 12, 11, 10.

   But now, having already converted to digital form, the solver should be particularly alert to note that the entire sequence is consecutive modulo 12 — in other words, in so-called clock arithmetic. So, the natural next term in the sequence is a 9. Finally, to complete the solution, we must reflect the digital representation of the 9 to get the term in the original sequence, which should therefore be a backwards 6, or ∂.

   Remark:

   Although this problem is quite tricky to 'backward engineer', it was arrived at in a very natural way. The idea was inspired by looking at an inverted clock face reflected in a mirror

2. What is the minimum number of different calendars it is necessary to have to cover every year? Justify your answer.

Solution:

The calendar for a year is completely determined by two pieces of information: (i) the first day of the year (or any other suitable reference day) and (ii) whether or not it is a leap year. From these two facts, the day on every date can be determined. This means that at most fourteen calendars are necessary to cover every year.

However, it is an easy exercise to confirm that any twenty-eight year period (not including the turn of a century, since this changes the pattern) will include examples of all fourteen combinations, and so all fourteen are necessary.

3. Decide whether or not the number

$$314154314155314156314157314158314159$$

is the square of an integer.

Solution:

This is a typical 'red-herring' problem. The largeness of the number, the repeated block of digits, and the similarity to the first six digits of π are all essentially irrelevant to solving the problem. In fact, the problem can be solved using only the fact that the number ends in the digits 59.

Recall that all squares of integers are congruent to zero or one modulo four. But $100k + 59 = 4(25k + 14) + 3 \equiv 3 \bmod 4$ for any integer k, and hence the given number is not congruent to zero or one modulo four, and hence is not the square of an integer.

Alternative Solution:

It is well known that a number is divisible by 3 if and only if the sum of its digits is divisible by 3, and similarly for divisibility by 9. Summing the digits of the given number, we get 123, which is a multiple of 3 but not of 9 — so the given number must be a multiple of 3 but not of 9, and hence cannot be a square.

4. Evaluate

$$\sum_{n=1}^{\infty} \tan^{-1}\left(\frac{1}{n^2 + n + 1}\right).$$

Solution:

In order to evaluate an infinite sum, we seek a telescoping form — i.e. we try to write the summand as $f(n+1) - f(n)$ for some function f. In order for the difference to be a tan inverse, we seek solutions of the form $\tan^{-1}(g(n+1)) - \tan^{-1}(g(n))$.

But now, if we set $a = \tan^{-1}(g(n+1))$ and $b = \tan^{-1}(g(n))$ in the usual tan difference rule

$$\tan(a-b) = \frac{\tan a - \tan b}{1 + \tan a \tan b}$$

we get

$$\tan^{-1}(g(n+1)) - \tan^{-1}(g(n)) = \tan^{-1}\left(\frac{g(n+1) - g(n)}{1 + g(n)g(n+1)}\right).$$

So, we wish to find a function $g(n)$ such that

$$\frac{g(n+1) - g(n)}{1 + g(n+1)g(n)} = \frac{1}{n^2 + n + 1}.$$

But now, if we simply equate the numerators and denominators in this expression, we get an obvious solution of $g(n) = n$. So we have $\tan^{-1}\left(\frac{1}{n^2+n+1}\right) = \tan^{-1}(n+1) - \tan^{-1}(n)$, and hence

$$\sum_{n=1}^{\infty} \tan^{-1}\left(\frac{1}{n^2+n+1}\right) = \lim_{k \to \infty} \left(\tan^{-1}(k) - \tan^{-1}(1)\right) = \frac{\pi}{2} - \frac{\pi}{4} = \frac{\pi}{4}.$$

5. If a, b, c, d, x, y, z and w are positive integers with $\frac{a}{b} > \frac{c}{d}$ and $\frac{x}{y} > \frac{z}{w}$, prove or disprove that

$$\frac{a+x}{b+y} > \frac{c+z}{d+w}.$$

Solution:

After attempting to find an algebraic proof of the inequality (which should not be forthcoming), we are motivated to attempt to prove that the conjecture is false.

Obviously $\frac{a+x}{b+y}$ lies between $\frac{a}{b}$ and $\frac{x}{y}$, and similarly $\frac{c+z}{d+w}$ lies between $\frac{c}{d}$ and $\frac{z}{w}$. This means that if the conjecture is to be false, we must have $\frac{c}{d} > \frac{x}{y}$ or $\frac{z}{w} > \frac{a}{b}$ (otherwise, the range of possible values of the left hand side is greater than the range of possible values for the right hand side). Without loss of generality, we may assume that

$$\frac{a}{b} > \frac{c}{d} > \frac{x}{y} > \frac{z}{w}.$$

In order to have the value of $\frac{a+x}{b+y}$ as small as possible, we need x and y to be as large as possible and a and b to be as small as possible (so that the weighting is strongly biased towards the lower value). Similarly, to make $\frac{c+z}{d+w}$ as large as possible, we need to make z and w small, and c and d large. To satisfy both of these conditions, and the inequalities, we seek solutions of the form $(a,b) = (1,2)$, $(c,d) = (n, 2n+1)$, $(x,y) = (m, 3m-1)$, $(z,w) = (1,3)$.

It is clear that, for sufficiently large m and n, we will have $\frac{a}{b} > \frac{c}{d} > \frac{x}{y} > \frac{z}{w}$. But also, as m gets large, $\frac{a+x}{b+y}$ will tend toward $\frac{1}{3}$, while as n gets large $\frac{c+z}{d+w}$ will tend towards $\frac{1}{2}$ — so for sufficiently large n and m, the inequality in the conjecture will be reversed.

In particular, with $(m,n) = (3,3)$, we get $\frac{1}{2} > \frac{3}{7}$ and $\frac{3}{8} > \frac{1}{3}$, but $\frac{1+3}{2+8} = \frac{2}{5} = \frac{3+1}{7+3}$, giving a case of equality. Also, with $(m,n) = (3,4)$, we get $\frac{1}{2} > \frac{4}{9}$ and $\frac{3}{8} > \frac{1}{3}$, but $\frac{1+3}{2+8} = \frac{2}{5} < \frac{5}{12} = \frac{4+1}{9+3}$, giving a case of reversal.

Remark:

Intuitively, this is a highly unusual result, which has some interesting manifestations in the real world. For instance, consider the hypothetical situation of two doctors running trials on two different drugs. The first doctor gives drug X to b people, of whom a recover, and gives drug Y to d people, of whom c recover. The second doctor gives drug X to y people, of whom x recover, and gives drug Y to w people, of whom z recover.

If each doctor found that drug X was more effective than drug Y (in the sense of curing a higher proportion of people treated — which is equivalent to the conditions $\frac{a}{b} > \frac{c}{d}$ and $\frac{x}{y} > \frac{z}{w}$), then we would expect that combining their results would provide an even stronger case for the superiority of drug X, when in fact drug Y may perform better when the results are combined (corresponding to $\frac{a+x}{b+y} < \frac{c+z}{d+w}$).

Another similar real-world application is in cricket, where if bowler A has a better average than bowler B in each of two innings, bowler B may still have a better average than bowler A in the match overall.

6. Evaluate the definite integral
$$\int_0^\pi \frac{x \sin x}{1 + \cos^2 x} dx.$$

Solution:

We note that, on the given range, $\sin x/(1 + \cos^2 x)$ is symmetric, so that we have
$$\frac{\sin t}{1 + \cos^2 t} = \frac{\sin(\pi - t)}{1 + \cos^2(\pi - t)}.$$

We can use this symmetry to rewrite the integral over the second half of the range as a similar integral over the first half of the range:
$$\int_{\frac{\pi}{2}}^\pi \frac{x \sin x}{1 + \cos^2 x} dx = \int_0^{\frac{\pi}{2}} \frac{(\pi - x) \sin(\pi - x)}{1 + \cos^2(\pi - x)} = \int_0^{\frac{\pi}{2}} \frac{(\pi - x) \sin x}{1 + \cos^2 x} dx.$$

Using this to simplify the entire integral, we get
$$\int_0^\pi \frac{x \sin x}{1 + \cos^2 x} dx = \int_0^{\frac{\pi}{2}} \left(\frac{x \sin x}{1 + \cos^2 x} + \frac{(\pi - x) \sin x}{1 + \cos^2 x} \right) dx$$
$$= \int_0^{\frac{\pi}{2}} \frac{\pi \sin x}{1 + \cos^2 x} dx.$$

Now that the integral is simplified to a purely trigonometric form, we can take a more traditional approach, using a change of variables. Let $u = \cos x$, so that we have $du = -\sin x \, dx$. Then we get
$$\int_0^\pi \frac{x \sin x}{1 + \cos^2 x} dx = \pi \int_0^{\frac{\pi}{2}} \frac{\sin x}{1 + \cos^2 x} dx = \pi \int_1^0 \frac{-du}{1 + u^2} = \pi \int_0^1 \frac{du}{1 + u^2}.$$

Now, we know that $\int \frac{dt}{1+t^2} = \tan^{-1} t$, so we can substitute this in to get a final result:
$$\int_0^\pi \frac{x \sin x}{1 + \cos^2 x} dx = \pi \int_0^1 \frac{du}{1 + u^2} = \pi \left(\tan^{-1} 1 - \tan^{-1} 0 \right) = \frac{\pi^2}{4}.$$

Remark:

This problem reappeared as question nine on the 2007 Superbrain, twenty years after it was first posed. It is not known how the standard of answer on the two occasions compared.

7. What are the minimum and maximum values of the function
$$f(x) = \frac{x^3}{3} + \frac{x^2}{2} - 2x$$
on the interval $[-3, 4]$?

Solution:

Since f is a polynomial, its extrema will occur either at the endpoints of the interval, or at critical points (i.e. zeros of the derivative). $f'(x) = x^2 + x - 2$, so $f'(x) = 0 \Leftrightarrow x = (-1 \pm 3)/2 = -2$ or 1. So the extreme values of f on the interval $[-3, 4]$ are in the set $f(\{-3, -2, 1, 4\}) = \{\frac{3}{2}, \frac{10}{3}, -\frac{7}{6}, \frac{64}{3}\}$, and hence the maximum value of the function on the given range is $\frac{64}{3}$, and its minimum value is $-\frac{7}{6}$.

8. A series of numbers, beginning with 17, is in both arithmetic and geometric progression. Find the sum of the first million terms of the series.

Solution:

Let r be the common ratio of the geometric progression, and d the common difference of the arithmetic progression. Then $17r = 17 + d$, and $17r^2 = 17 + 2d$. Eliminating d from the two equations, we get $17r^2 - 34r + 17 = 17(r-1)^2 = 0$, and hence $r = 1$. This also gives $d = 0$, and from either one of these we can deduce that the sequence is constant — that is, every term is 17.

The sum of the first million terms is therefore simply 17000000.

9. ABC is a triangle. Show how to construct a square $XYZW$ such that A lies on the line XY, B lies on the line YZ and C lies on the line XW.

Solution:

We must arrange that A is on the side-line of the square which is opposite the side-line which is not required to contain a vertex of the triangle. In the diagram we supply, it so happens that B is obtuse. The reader should reflect and convince herself that the procedure would work for any triangle. Let $Y = A$. We then construct a line through C parallel to AB, and a line through A perpendicular to AB, and let the intersection of these be X.

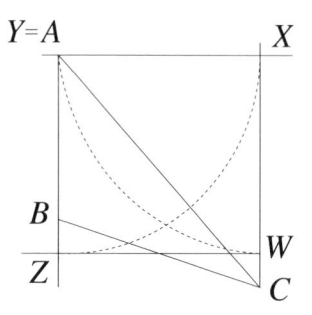

We then centre the compasses at X with radius $|XY|$, and let the intersection of this arc with the line CX be W. Finally, we centre the compasses on Y with radius $|XY|$, and let the intersection of this arc with line AB be Z. Be careful to choose W and Z on the same side of the line XY.

10. Find all integers x and y which satisfy the equation $x^2 - 3y^2 = 17$.

    Solution:

    The fact that a 3 appears in a Diophantine equation with squares of integers should remind the problem solver of the useful fact that no square of an integer is congruent to two modulo three. This provides an immediate solution to this problem, since

    $$x^2 = 3y^2 + 17 = 3(y^2 + 5) + 2 \equiv 2 \bmod 3,$$

    and so there are no integer solutions for x.

Fifth Superbrain, 1988 — Solutions

1. What, in your opinion is the next term in the sequence F, G, J, L, P, Q, and why?

   Solution:

   The first thing to note about this sequence is that each of its terms are letters, and that they appear in alphabetical order. This suggests that the sequence may be simply the set of all letters with a certain property listed in order. To help us to see what this property is, we might write out the set of letters which have it side by side with the set of letters which do not:

   $$\text{F, G, J, L, P, Q, ...}$$
   $$\text{A, B, C, D, E, H, I, K, M, N, O, ...}$$

   Although it is hard to give a systematic means to determine the property being sought, intuitively the solver may well feel that the letters above are the more 'irregular' ones in terms of shape (unfortunately, if this intuition does not materialise, then the problem is likely to be intractable).

   Having decided this, the solver might try to apply a more rigorous notion of 'regularity' to the problem — and one such mathematical rigour presents itself in the form of symmetry: all of the letters in the second row have symmetry (either axial in most cases, or rotational in the case of 'N').

   So we might conclude that the given sequence is the set of symmetry-free letters in alphabetical order, and therefore that the next term in the sequence should be 'R'.

2. Find all real numbers a, b and c such that

   $$a + b + c = 0 = \frac{1}{a} + \frac{1}{b} + \frac{1}{c}.$$

   Solution:

   For $\frac{1}{a}, \frac{1}{b}$ and $\frac{1}{c}$ to be defined, each of a, b, c must be non-zero. So we can multiply through by abc in the second equation to get $ab + bc + ca = 0$.

   But now, $a + b + c = 0$, so

   $$0 = (a + b + c)^2 = a^2 + b^2 + c^2 + 2(ab + bc + ca) = a^2 + b^2 + c^2.$$

But $a^2 + b^2 + c^2 \geq 0$, with equality if and only if $a = b = c = 0$, which contradicts the fact that none of the numbers are zero. To avoid contradiction, there must be no real numbers which satisfy the given equations.

3. If n is a natural number, prove that
$$n^n \geq 1.3.5.7 \ldots (2n-1).$$

Solution:

Let $P = 1.3.5.7 \ldots (2n-1)$. Then we have
$$P^2 = [1.(2n-1)] \ldots [(2n-1).1] = \prod_{k=1}^{n}(2k-1).(2n-2k+1).$$

But $(2k-1)(2n-2k+1) \leq n^2$, since $0 \leq (n-2k+1)^2$ (or by the arithmetic-geometric mean inequality), so
$$P^2 \leq \prod_{k=1}^{n} n^2 = n^{2n},$$

and taking square roots (since both P and n^n are positive), we get $P \leq n^n$ as required.

Alternative Solution:

A more direct solution can be obtained by applying the n variable arithmetic-geometric mean inequality right from the start. The arithmetic mean of the n numbers on the right hand side is n, and so their product is at most n^n.

4. Find all real numbers x, y such that $(x+iy)^2 = -i$, where $i^2 = -1$.

Solution:

Expanding the square in the given equation, we get $-i = x^2 - y^2 + 2xyi$. Equating real and imaginary parts gives $x^2 - y^2 = 0$ and $2xy = -1$. From the first equation, we get that $y = \pm x$. But $y = x$ gives $-1 = 2xy = 2x^2 \geq 0$, a contradiction. So $y = -x$.

Substituting this into the second equation, we get $2x^2 = 1$, which gives $x = -y = \pm\frac{1}{\sqrt{2}}$. So we get two possible solutions:
$$(x, y) \in \left\{\left(\frac{1}{\sqrt{2}}, -\frac{1}{\sqrt{2}}\right), \left(-\frac{1}{\sqrt{2}}, \frac{1}{\sqrt{2}}\right)\right\}.$$

5. Evaluate
$$\int \frac{\sqrt{x}}{1+\sqrt[3]{x}}dx.$$

Solution:

To evaluate this integral, we would like to remove the two surds, and create a more straightforward rational integrand.

The square and cube roots of x are both integer powers of the sixth root, so we try the substitution $u = \sqrt[6]{x}$.

This gives $x = u^6$, and hence $dx = 6u^5 du$, so we get

$$\int \frac{\sqrt{x}}{1+\sqrt[3]{x}}dx = \int \frac{(u^3)(6u^5)}{1+u^2}du = 6\int \frac{u^8}{1+u^2}du$$

$$= 6\int \left(u^6 - u^4 + u^2 - 1 + \frac{1}{1+u^2}\right)du$$

$$= \frac{6u^7}{7} - \frac{6u^5}{5} + 2u^3 - 6u + 6\tan^{-1}(u) + C$$

$$= \frac{6}{7}x^{\frac{7}{6}} - \frac{6}{5}x^{\frac{5}{6}} + 2\sqrt{x} - 6\sqrt[6]{x} + 6\tan^{-1}\left(\sqrt[6]{x}\right) + C,$$

where C is an arbitrary constant.

6. Investigate
$$\lim_{x\to 0} \frac{\sin\frac{1}{x}}{\sin\frac{1}{x}}.$$

Solution:

This question appears to be trivial — to the point where it is often mistaken for a misprint. Obviously anything divided by itself is equal to one, and so the limit must be one. In fact, the question is designed to exploit a difference between the rigorous definition of a limit and the intuitive idea.

We observe that $\frac{\sin\frac{1}{x}}{\sin\frac{1}{x}}$ is not defined when $\sin\frac{1}{x} = 0$, or when $x = \frac{1}{n\pi}$ for any non-zero integer n (since $\frac{0}{0}$ is not defined). Moreover, the limit of $f(x)$ as x tends towards x_0 exists only if $f(x)$ is defined on some neighbourhood containing x_0 (except possibly at x_0).

But now, $\lim_{n\to\infty} \frac{1}{n\pi} = 0$, so any neighbourhood of 0 contains infinitely many points of the form $\frac{1}{n\pi}$ for n a non-zero integer, and hence $\frac{\sin\frac{1}{x}}{\sin\frac{1}{x}}$ is not defined on any neighbourhood of zero. So $\lim_{x\to 0} \frac{\sin\frac{1}{x}}{\sin\frac{1}{x}}$ does not exist.

7. Find the value of $\frac{1}{3} + \frac{2}{5} + \frac{4}{17} + \frac{8}{257} + \ldots = \sum_{n=0}^{\infty} \frac{2^n}{2^{2^n} + 1}$.

Solution:

$$\frac{1}{3} + \frac{2}{5} = \frac{11}{15} \qquad \frac{11}{15} + \frac{4}{17} = \frac{247}{255} \qquad \frac{247}{255} + \frac{8}{257} = \frac{65519}{65535}.$$

Adding up the first few terms of the series and observing that the terms of the sequence appear to be converging rapidly to zero suggests that the limit being sought is one. This motivates us to consider the difference between the first few partial sums and one. These are

$$\frac{2}{3}, \qquad \frac{4}{15}, \qquad \frac{8}{255} \quad \text{and} \quad \frac{16}{65535}.$$

The similarity to the terms of the series is obvious, and leads us to the conjecture that the sum of the first n terms is $1 - \frac{2^n}{2^{2^n} - 1}$. We prove this conjecture by induction.

For $n = 1$, the conjecture is clearly true. So we assume that the conjecture is true for $n = k$ — that is, that $\sum_{n=0}^{k-1} \frac{2^n}{2^{2^n} + 1} = 1 - \frac{2^k}{2^{2^k} - 1}$. Then we have

$$\sum_{n=0}^{k} \frac{2^n}{2^{2^n} + 1} = 1 - \frac{2^k}{2^{2^k} - 1} + \frac{2^k}{2^{2^k} + 1} = 1 - \frac{2^{k+1}}{2^{2^{k+1}} - 1},$$

and so the conjecture is proved for $n = k + 1$. By induction, the conjecture holds for all positive integers.

So $\sum_{n=0}^{\infty} \frac{2^n}{2^{2^n} + 1} = \lim_{k\to\infty} \left(1 - \frac{2^k}{2^{2^k} - 1}\right) = \lim_{m\to\infty} \left(1 - \frac{m}{m^2 - 1}\right) = 1$.

8. A plane curve is given by

$$y = ax^3 + bx^2 + cx + d,$$

for $a \neq 0$. Find conditions on a, b, c and d which ensure that the curve has a maximum point and a minimum point locally. Under these conditions show that the curve has a point of inflection that lies midway between the local maximum point and the local minimum point.

Solution:

Since the curve in question is a cubic curve, it will have a maximum point and a minimum point locally if and only if the value of its derivative changes sign. But $\frac{dy}{dx} = 3ax^2 + 2bx + c$ is a quadratic, so its value changes sign if and only if the discriminant is positive, that is if and only if $(2b)^2 - 4(3a)(c) > 0$. So, the curve has a maximum point and a minimum point locally if and only if $b^2 > 3ac$.

If this condition is satisfied, then the two extremal points will be the two roots of the derivative, the average of which will be $x = -\frac{b}{3a}$ (since the average of the two roots of $px^2 + qx + r = 0$ is $-\frac{q}{2p}$). But the point of inflection occurs when the second derivative is zero, or when $\frac{d^2y}{dx^2} = 6ax + 2b = 0$. But this is precisely at $x = -\frac{b}{3a}$, which is half way between the two roots as required.

9. A circle is inscribed in a quadrant of a circle whose radius is 10cm. Find the radius of the inscribed circle.

Solution:

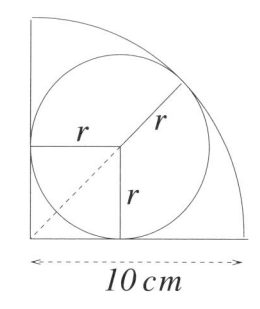

Consider the radius of the larger circle passing through the centre of the smaller one. This is composed of two parts: the part joining the two centres, and the part from the centre of the smaller circle to the circumference of the two circles.

Clearly the second part is a radius of the smaller circle. We assign this a length r. But then, the first part is the diagonal of a square of side r (see diagram), so that it has length $\sqrt{2}r$. Therefore the radius of the larger circle has length $r(1 + \sqrt{2}) = 10cm$.

Rearranging, we get $r = 10(\sqrt{2} - 1)cm$.

10. Prove, using vectors, that the angle in a semi-circle is a right angle.

   Solution:

   Seeing a requirement to prove that something is a right angle using vectors should immediately bring the dot product to mind. The angle between two vectors (of non-zero length) is a right angle if and only if their dot product is zero.

   Let A, B and C be points on a circle with centre O and radius r, and let A and C be diametrically opposite. Then we have $\vec{BC} = \vec{OC} - \vec{OB}$, while $\vec{AB} = \vec{OB} - \vec{OA} = \vec{OB} + \vec{OC}$. So, we get

   $$\vec{AB} \cdot \vec{BC} = (\vec{OC} + \vec{OB}) \cdot (\vec{OC} - \vec{OB}) = |\vec{OC}|^2 - |\vec{OB}|^2 = r^2 - r^2 = 0.$$

   So the dot product of $\vec{AB}$ and $\vec{BC}$ is zero, and hence the angle between them, which is also the angle in the semi-circle, $\angle ABC$, is a right angle as required.

   Alternative Solution:

   We may suppose that the circle is centred at $(0,0)$, and that it has a radius of $r > 0$. Let $A = (-r, 0)$, $B = (r\cos\theta, r\sin\theta)$ and $C = (r, 0)$. Then AC is a diameter, and B is a point in the semi-circle as before.

   Now we have $\vec{AB} = (r\cos\theta + r, r\sin\theta)$, $\vec{BC} = (r - r\cos\theta, -r\sin\theta)$, so $\vec{AB} \cdot \vec{BC} = r^2(1 - \cos^2\theta - \sin^2\theta) = 0$, and the result follows as before.

Sixth Superbrain, 1989 — Solutions

1. Given the coordinates of three distinct points A, B, C in the plane, give ten essentially different methods of deciding whether or not A, B, and C lie on the same line.

   Solution:

   (a) Determine the lengths $|AB|$, $|BC|$ and $|CA|$. If the greatest of these is the sum of the other two, then the three are collinear. Otherwise, they are not. (Since collinearity is the only case of equality in the triangle inequality).

   (b) Substitute the coordinates of each of the three points into the equation of a circle, $(x-a)^2+(y-b)^2 = r^2$. If the resulting simultaneous equations can be solved for a, b and r then the three points are not collinear, but, if they cannot, then the three points are collinear (since there exists a circle passing through three points if and only if they are not collinear).

   (c) Determine the slope of the line segments AB and BC. The three points are collinear if and only if these are equal (or both undefined, corresponding to vertical line-segments).

   (d) Show that no parabola passes through all three points.

   (e) Try to solve the vector equation $t\vec{A} + (1-t)\vec{B} = \vec{C}$. If it has a solution for t, then the points are collinear, otherwise they are not.

   (f) Determine the sidelengths of $\triangle ABC$, and hence its area. The three points are collinear if and only if the area is zero.

   (g) Determine the sidelengths of $\triangle ABC$ and use the cosine rule to evaluate $|\cos \angle ABC|$. The three points are collinear if and only if the value of this is 1.

   (h) The three points are non-collinear if and only if $\{\overrightarrow{AB}, \overrightarrow{BC}\}$ is a basis for the set of all vectors in the plane.

   (i) Translate all three points so that A is the origin, and then rotate so that B lies of the x−axis. The three are collinear if and only if C then also lies on the x−axis.

   (j) If the points have co-ordinates (x_1, y_1), (x_2, y_2) and (x_3, y_3) then they are collinear if and only if the determinant
   $$\begin{vmatrix} 1 & 1 & 1 \\ x_1 & x_2 & x_3 \\ y_1 & y_2 & y_3 \end{vmatrix}$$

is zero — that is, if and only if

$$x_1 y_2 + x_2 y_3 + x_3 y_1 - x_1 y_3 - x_2 y_1 - x_3 y_2 = 0.$$

Remark:

The authors leave to the reader the question of determining which of the above methods should be considered the 'best'.

2. Given n consecutive positive integers, show that $n!$ is a factor of their product.

Solution:

The product of n consecutive positive integers is $\frac{(n+k)!}{k!}$, where $k+1$ is the smallest of the numbers.

But $\frac{(n+k)!}{k! n!} = \binom{n+k}{k}$, the number of k-element subsets of an $n+k$-element set, which must be an integer. So $\frac{(n+k)!}{k!} = \binom{n+k}{k} n!$, which is a multiple of $n!$ as required.

3. Evaluate the definite integral

$$\int_0^{\frac{\pi}{2}} \frac{\sin^{17} \theta}{\cos^{17} \theta + \sin^{17} \theta} d\theta.$$

Solution:

The key to evaluating this integral is to take advantage of the limits of integration (the same indefinite integral would be much harder). We use the fact that sine and cosine are mutually symmetric on $\left(0, \frac{\pi}{2}\right)$ and the rule $\int_a^b f(x) dx = \int_a^b f(a+b-x) dx$ to do this.

If

$$I = \int_0^{\frac{\pi}{2}} \frac{\sin^{17} \theta}{\cos^{17} \theta + \sin^{17} \theta} d\theta,$$

then

$$I = \int_0^{\frac{\pi}{2}} \frac{\sin^{17} \left(\frac{\pi}{2} - \theta\right)}{\cos^{17} \left(\frac{\pi}{2} - \theta\right) + \sin^{17} \left(\frac{\pi}{2} - \theta\right)} d\theta = \int_0^{\frac{\pi}{2}} \frac{\cos^{17} \theta}{\cos^{17} \theta + \sin^{17} \theta} d\theta.$$

Averaging the two expressions for I, we get

$$I = \frac{1}{2} \int_0^{\frac{\pi}{2}} \frac{\cos^{17} \theta + \sin^{17} \theta}{\cos^{17} \theta + \sin^{17} \theta} d\theta = \frac{1}{2} \int_0^{\frac{\pi}{2}} d\theta = \frac{\pi}{4}.$$

Remark:

The same concept appears in a more intimidating guise as question 5 on the 2000 paper — although to the experienced solver, the larger numbers there may make the approach more obvious.

4. Find any three positive whole numbers a, b, and c which satisfy the equation $a^3 + b^4 = c^5$.

   Solution:

   To find a simple solution to this problem, we must find two perfect integer powers whose sum is also a perfect integer power. The most simple general case of this is $2^n + 2^n = 2^{n+1}$, and indeed, if we can find positive integers a, b, c and n such that $a^3 = 2^n$, $b^4 = 2^n$ and $c^5 = 2^{n+1}$, then we will be done.

   But if a power of an integer is a power of two, then the only possible prime factor of that integer is two, and so it must itself be a power of two. So it is sufficient to find positive integers α, β and γ such that $a = 2^\alpha$, $b = 2^\beta$ and $c = 2^\gamma$. So we wish to find positive integers α, β, γ and n, with $3\alpha = 4\beta = 5\gamma - 1 = n$.

   $3|n$ and $4|n$, so $12|n$, and $n \equiv -1 \bmod 5$. By inspection, $n = 24$ satisfies both requirements (and by the Chinese Remainder Theorem, it is the only solution modulo $3.4.5 = 60$). So, we have $\alpha = \frac{n}{3} = 8$, $\beta = \frac{n}{4} = 6$ and $\gamma = \frac{n+1}{5} = 5$.

   Substituting these into our expressions for a, b and c, we have a solution to the original problem:

   $$a = 2^8, \quad b = 2^6, \quad c = 2^5.$$

   Remark:

   In fact, the solution arrived at by this method is the smallest solution to the given equation.

5. Given five distinct real numbers x_1, x_2, x_3, x_4, x_5, prove that at least two of them satisfy the inequality

   $$0 < \frac{x_i - x_j}{x_i x_j + 1} < 1.$$

Solution:

The expression in the inequality is strongly reminiscent of the formula for the tangent of the difference between two angles:

$$\tan(a - b) = \frac{\tan a - \tan b}{1 + \tan a \tan b}.$$

To take advantage of this, let $\theta_i = \tan^{-1} x_i$ for $i = 1, 2, 3, 4, 5$. Then we get $\tan(\theta_i - \theta_j) = \frac{x_i - x_j}{x_i x_j + 1}$.

Now, by the definition of the inverse tangent, $\theta_i \in \left(-\frac{\pi}{2}, \frac{\pi}{2}\right)$ for each i. So by the pigeonhole principle, at least one of the four ranges $\left(-\frac{\pi}{2}, -\frac{\pi}{4}\right]$, $\left(-\frac{\pi}{4}, 0\right]$, $\left(0, \frac{\pi}{4}\right]$ and $\left(\frac{\pi}{4}, \frac{\pi}{2}\right)$ must contain at least two of the θ_i.

But since no two angles in any of these ranges differ by $\pi/4$ or more, and no two are the same, there must be some i, j for which $\theta_i - \theta_j \in \left(0, \frac{\pi}{4}\right)$, and hence

$$\tan(\theta_i - \theta_j) = \frac{x_i - x_j}{x_i x_j + 1} \in (0, 1)$$

as required.

6. Let $ABCD$ be a trapezium with AB parallel to DC. The diagonals AC and BD intersect at R and the line MRN is parallel to AB, where M lies on AD and N lies on BC. Prove that $|MR| = |RN|$.

Solution:

Let $\Delta(XYZ)$ be the area of the triangle with vertices at X, Y and Z, and let h be the distance between the lines AB and DC.

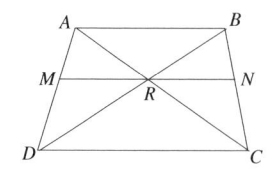

$\Delta(ACD) = \Delta(BCD)$, since the two triangles have base $|DC|$ and height h. So $\Delta(ARD) = \Delta(BRC)$, by removing the area common to the two. But now,

$$|MR| = \frac{2\Delta(ARD)}{h} = \frac{2\Delta(BRC)}{h} = |RN|.$$

Alternative Solution:

The reader familiar with affine transformations could apply a shear to get an isosceles trapezium. This clearly makes the two lengths equal. But affine transformations preserve ratios of lengths on a line, so this must hold for the original trapezium as well.

7. Evaluate $\dfrac{1}{3} + \dfrac{1}{3} + \dfrac{1}{5} + \dfrac{4}{45} + \ldots = \displaystyle\sum_{n=1}^{\infty} \dfrac{n 2^n}{(n+2)!}$.

Solution:

$$\dfrac{1}{3} + \dfrac{1}{3} = \dfrac{2}{3} \qquad \dfrac{2}{3} + \dfrac{1}{5} = \dfrac{13}{15} \qquad \dfrac{13}{15} + \dfrac{4}{45} = \dfrac{43}{45} \qquad \dfrac{43}{45} + \dfrac{2}{63} = \dfrac{311}{315}.$$

Evaluating the first few partial sums and observing that the terms appear to be converging rapidly to zero suggests that the sum converges to one. To test this possibility, we consider the differences between the first few partial sums and one. These are

$$\dfrac{2}{3} \qquad \dfrac{1}{3} \qquad \dfrac{2}{15} \qquad \dfrac{2}{45} \qquad \dfrac{4}{315}.$$

The pattern is hard to spot in this form, because common factors have been cancelled from above and below in adding up the terms. However, if we write and add the fractions with no cancellation (so that the sum of the first n terms has a denominator of $(n+2)!$), the pattern becomes more obvious:

$$\dfrac{4}{6} \qquad \dfrac{8}{24} \qquad \dfrac{16}{120} \qquad \dfrac{32}{720} \qquad \dfrac{64}{5040}.$$

This leads us to conjecture that the sum of the first n terms of the sequence is $1 - \dfrac{2^{n+1}}{(n+2)!}$. We prove this conjecture by induction. For $n=1$, the conjecture is true, so we assume that it is true for $n = k-1$ — that is, that $\displaystyle\sum_{n=1}^{k-1} \dfrac{n 2^n}{(n+2)!} = 1 - \dfrac{2^k}{(k+1)!}$. Then

$$\sum_{n=1}^{k} \dfrac{n 2^n}{(n+2)!} = 1 - \dfrac{2^k}{(k+1)!} + \dfrac{k 2^k}{(k+2)!} = 1 - \dfrac{2^{k+1}}{(k+2)!},$$

and so the conjecture is true for $n = k$. By induction, we have $\displaystyle\sum_{n=1}^{k} \dfrac{n 2^n}{(n+2)!} = 1 - \dfrac{2^{k+1}}{(k+2)!}$ for all positive integers k.

So, $\displaystyle\sum_{n=1}^{\infty} \dfrac{n 2^n}{(n+2)!} = \lim_{k\to\infty}\left(1 - \dfrac{2^k}{(k+1)!}\right) = 1 - \lim_{k\to\infty} \dfrac{2^k}{(k+1)!}.$

But $0 < \dfrac{2^k}{(k+1)!} = \prod_{n=2}^{k+1}\dfrac{2}{n} < \prod_{n=4}^{k+1}\dfrac{2}{n} < \dfrac{1}{2^{k-2}}$ (for $k \geq 3$), and $\lim_{k\to\infty}\dfrac{1}{2^{k-2}} = 0$. So, by the squeeze (pinching) lemma, $\lim_{k\to\infty}\dfrac{2^k}{(k+1)!} = 0$. Putting this into the above expression for the sum, we get

$$\sum_{n=1}^{\infty}\dfrac{n2^n}{(n+2)!} = 1,$$

exactly as we had conjectured.

Alternative Solution:

It is possible to solve this problem by generalizing it. Let

$$f(x) = \sum_{n=0}^{\infty}\dfrac{nx^{n+2}}{(n+2)!}.$$

Then differentiating twice gives

$$f''(x) = \sum_{n=1}^{\infty}\dfrac{x^n}{(n-1)!} = \sum_{n=0}^{\infty}\dfrac{x^{n+1}}{n!} = xe^x.$$

But now, it is easy so see that each term in the expansion of $f(x)$ and $f'(x)$ is zero for $x = 0$, and so $f'(0) = f(0) = 0$. Using this information and the general expression for the double derivative, we can solve for $f(x)$:

$$f(x) = \int_0^x\int_0^t se^s\,ds\,dt = \int_0^x\left[(t-1)e^t + 1\right]dt = (x-2)e^x + x + 2.$$

To complete the solution, we need only note that the desired sum is

$$\sum_{n=0}^{\infty}\dfrac{n2^n}{(n+2)!} = \dfrac{1}{4}\sum_{n=0}^{\infty}\dfrac{n2^{n+2}}{(n+2)!} = \dfrac{1}{4}f(2) = \dfrac{1}{4}\left((2-2)e^2 + 2 + 2\right) = 1.$$

8. The lengths of the sides of a triangle are in geometric progression with common ratio r. Prove that $1/\phi < r < \phi$, where $\phi = (1+\sqrt{5})/2$.

Solution:

Let the three sides be a, ar and ar^2. Obviously $a, r > 0$, and by the triangle inequality $a + ar > ar^2$ and $ar^2 + ar > a$. So $r^2 - r - 1 < 0$ and $r^2 + r - 1 > 0$. Since both of the quadratics in r have positive coefficients of r^2, they will be negative between their two roots, and positive outside this range. So r must lie between the roots of $x^2 - x - 1 = 0$, and outside the roots of $x^2 + x - 1 = 0$.

The polynomial $x^2 - x - 1$ has roots at $x = (1 \pm \sqrt{5})/2$, so

$$\frac{1-\sqrt{5}}{2} < r < \frac{1+\sqrt{5}}{2} = \phi.$$

The polynomial $x^2 + x - 1$ has roots at $x = (-1 \pm \sqrt{5})/2$, so $r < (-1-\sqrt{5})/2 < 0$ (which is impossible), or $r > (\sqrt{5}-1)/2$. But

$$\frac{\sqrt{5}-1}{2} = \frac{2}{\sqrt{5}+1} = \frac{1}{\phi},$$

so $1/\phi < r$.

Combining the two inequalities, we get $1/\phi < r < \phi$ as required.

9. If a, b, c are numbers such that

$$\frac{1}{a} + \frac{1}{b} + \frac{1}{c} = \frac{1}{a+b+c},$$

prove that

$$\frac{1}{a^{23}} + \frac{1}{b^{23}} + \frac{1}{c^{23}} = \frac{1}{a^{23}+b^{23}+c^{23}}.$$

Solution:

Multiplying the first equation through by $abc(a+b+c)$, we get that $(ab+bc+ca)(a+b+c) = abc$. Expanding and gathering terms to one side, this gives

$$0 = a^2b + ab^2 + b^2c + bc^2 + c^2a + ca^2 + 2abc = (a+b)(b+c)(c+a).$$

Now, since $abc(a+b+c)$ cannot be zero (otherwise one or more of the fractions would not be defined), it follows that

$$\frac{1}{a} + \frac{1}{b} + \frac{1}{c} = \frac{1}{a+b+c}$$

if, and only if, $a+b=0$ or $b+c=0$ or $c+a=0$.

But $a+b=0 \Leftrightarrow a=-b \Leftrightarrow a^{23}=-b^{23} \Leftrightarrow a^{23}+b^{23}=0$, and similarly for $b+c=0$ and $c+a=0$.

Combining all of these, it follows that $\dfrac{1}{a} + \dfrac{1}{b} + \dfrac{1}{c} = \dfrac{1}{a+b+c}$
$\Leftrightarrow (a+b)(b+c)(c+a) = 0 \Leftrightarrow (a^{23} + b^{23})(b^{23} + c^{23})(c^{23} + a^{23}) = 0$
$\Leftrightarrow \dfrac{1}{a^{23}} + \dfrac{1}{b^{23}} + \dfrac{1}{c^{23}} = \dfrac{1}{a^{23} + b^{23} + c^{23}}$, and the result is established.

Remark:

It is easy to see from the method used that the 23 in this problem could have been replaced by any odd number.

10. Each side of a right-angled triangle has an integer length, and the its perimeter is twice its area. Find the length of each side.

Solution:

Let the perpendicular sides have lengths a and b. Then the perimeter is $a + b + \sqrt{a^2 + b^2}$, the area is $ab/2$, and the equation which we must solve is $a + b + \sqrt{a^2 + b^2} = ab$, or $a^2 + b^2 = (ab - a - b)^2$.

Expanding the brackets and simplifying, we get $ab(ab - 2a - 2b + 2) = 0$. But $a, b > 0$, so we get $ab - 2a - 2b + 2 = 0$, which can be partially factored as $(a - 2)(b - 2) = 2$.

But since 2 can be written as a product of two integers in exactly one way, we have $\{a - 2, b - 2\} = \{1, 2\}$, and hence $\{a, b\} = \{3, 4\}$. So the only possible triangle is the one with sidelengths 3, 4 and 5.

Alternative Solution:

This problem can be approached in a different manner by using the well-known formula for integer Pythagorean triples

$$(a, b, c) = \left(2mnr, r(m^2 - n^2), r(m^2 + n^2)\right),$$

where a, b and c are the sidelengths (c being the hypotenuse), and m, n and r are positive integers with $m > n$, and where m and n are coprime and of different parity (i.e. one is odd and the other is even).

Using this substitution, the perimeter becomes $2mr(m + n)$, the area becomes $mnr^2(m^2 - n^2)$ and the equation to be solved is

$$2mr(m + n) = 2mnr^2(m^2 - n^2).$$

Cancelling common factors (which are all greater than zero), we get $nr(m - n) = 1$, and so $n = r = m - n = 1$, or $(m, n, r) = (2, 1, 1)$. It is easy to see that this gives the same triangle as the first solution.

Seventh Superbrain, 1990 — Solutions

1. A car is driven from A to B at an average speed of $40km$ per hour and is then driven from B to A along the same path at an average speed of $60km$ per hour. What is the average speed of the car for the entire journey?

   Solution:

   Let $|AB|$ be the distance (in kilometres) from A to B along the path travelled. Then the time taken for the outward journey is $\frac{|AB|}{40}$ hours, and the time taken for the return journey is $\frac{|AB|}{60}$ hours. So the total travelling time is $\frac{|AB|}{40} + \frac{|AB|}{60}$ hours, while the total distance travelled is $2|AB|$.

   Combining the total time and distance of the journey, we get an average journey speed of
   $$\frac{2|AB|}{\frac{|AB|}{40} + \frac{|AB|}{60}} = 48km \text{ per hour.}$$

   Remark:

   This question plays on the intuitive assumption that all averages must be arithmetic means — it seems obvious that the answer will be $50km$ per hour, rather than 48. In fact, in general the average speed over several journeys of the same distance is the harmonic mean of the speeds for each journey (that is, the inverse of the average of the inverses). As with many counter-intuitive questions, it is necessary to carefully calculate the solution rather than taking the apparent shortcut.

2. If n, x, y and z are all positive integers, find all solutions of the equation
   $$n^x + n^y = n^z.$$

   Solution:

   Since $n^m > 0$ for all positive integers m, we have $n^z = n^x + n^y > n^x$, and similarly $n^z > n^y$. But n^a is an increasing function of a, so we must have $z > x, y$. But since x, y and z are all integers, we must have $z - 1 \geq x, y$.

   This gives $n^z = n^x + n^y \leq 2n^{z-1}$, which gives $2 \geq n$. Obviously $n = 1$ gives $1 + 1 = 1$, which is false, so we must have $n = 2$. Finally, equality occurs in $2 \geq n$ only if equality occurs in $z - 1 \geq x$ and $z - 1 \geq y$. So
   $$(n, x, y, z) = (2, m, m, m+1)$$

for some positive integer m. Conversely, it is easy to see that every quadruple of this form satisfies the equation given, so that this is exactly the solution set desired.

Alternative Solution:

Having established that $z > x, y$, we could then proceed by dividing across by n^z, to get
$$\frac{1}{n^{z-x}} + \frac{1}{n^{z-y}} = 1.$$
But now, it is known (and very easy to prove in any case) that the only integers a and b which satisfy $\frac{1}{a} + \frac{1}{b} = 1$ are $a = b = 2$. So, we have $n^{z-x} = n^{z-y} = 2$. The rest follows easily from here.

3. Prove that
$$\sin 18° = \frac{\sqrt{5} - 1}{2}.$$

Solution:

By de Moivre's Theorem, $\sin 5\theta =$
$$\sin^5 \theta - 10 \sin^3 \theta \cos^2 \theta + 5 \sin \theta \cos^4 \theta$$
Now since $\cos^2 \theta = 1 - \sin^2 \theta$, this can be written as
$$16 \sin^5 \theta - 20 \sin^3 \theta + 5 \sin \theta.$$
If we let $\theta = (4n+1)\pi/10$ for some integer n, and $s = \sin \theta$, we get
$$16s^5 - 20s^3 + 5s - 1 = 0$$
(since $\sin 5\theta = \sin \frac{\pi}{2} = 1$). Now, the five roots of this equation must be the five different possible values of $\sin \theta$. But since one of these values is $\sin \frac{\pi}{2} = 1$, one of the roots must be $s = 1$. Moreover, $\sin \frac{\pi}{10} = \sin \frac{9\pi}{10}$, and $\sin \frac{13\pi}{10} = \sin \frac{17\pi}{10}$, so that the other four roots are in repeated pairs. This allows us to write $16s^5 - 20s^3 + 5s - 1 = (s-1)(as^2 + bs + c)^2$, and solving for a, b and c we get $16s^5 - 20s^3 + 5s - 1 = (s-1)(4s^2 + 2s - 1)^2$. So $\sin 18°$ must be a root of the quadratic $4s^2 + 2s - 1 = 0$. Solving this equation, we get $\sin 18° \in \left\{ \frac{\sqrt{5}-1}{4}, \frac{-\sqrt{5}-1}{4} \right\}$.

However, sine is an increasing function on $\left(0, \frac{\pi}{2}\right)$, so we must have $\frac{-\sqrt{5}-1}{4} < 0 = \sin 0° < \sin 18°$. So $\sin 18° = \frac{\sqrt{5}-1}{4}$, as required.

Alternative Solution:

Instead of using $5\theta = 2n\pi + \pi/2$ to get $\sin 5\theta = 1$, we can also use it to get $\sin 3\theta = \cos(\pi/2 - 3\theta) = \cos 2\theta$. A standard trigonometric expansion, and again setting $\sin\theta = s$ gives

$$3s - 4s^3 = 1 - 2s^2.$$

Now, gathering all terms to one side, we get a cubic in s. However, we know automatically that $\theta = \pi/2$ satisfies the given equation, and hence $s = \sin\theta = 1$ must be a root. Factoring this out, we get

$$(s-1)(4s^2 + 2s - 1) = 0,$$

and we can proceed as above to complete the solution.

4. At what time between 1pm and 2pm do the minute hand and the hour hand of a clock coincide exactly?

Solution:

The minute hand travels at a rate of one revolution per hour, while the second hand travels at a rate of $1/12$ revolutions per hour. At 1pm, the hour hand is $1/12$ revolutions ahead of the minute hand, and we must determine exactly how long it takes for the minute hand to catch up.

The minute hand catches up with the hour hand at a rate of $1 - 1/12 = 11/12$ revolutions per hour, and so catching up a lead of $1/12$ revolutions will take exactly $(1/12)/(11/12) = 1/11$ hours.

So, the minute hand and the hour hand coincide exactly at $1/11$ hours past 1pm, or at $13:05:27\frac{3}{11}$.

5. Solve the equation

$$30x^4 + 61x^3 + 90x^2 + 61x + 30 = 0.$$

Solution:

The symmetry of the coefficients in this equation is striking, and indeed, it is the key to solving the equation. Since $x = 0$ is not a solution of the equation, we may divide by x^2, so that the powers of x become more symmetric as well:

$$30x^2 + 61x + 90 + 61x^{-1} + 30x^{-2} = 0.$$

Now, to take advantage of the symmetry, we replace x with a more symmetric variable. In particular, if we let $u = x + 1/x$, then we get $30u^2 + 61u + 30 = 0$. Solving this quadratic in u, we get

$$u = \frac{-61 \pm \sqrt{121}}{60} = -1 \text{ or } -\frac{6}{5}.$$

But now, $x^2 - ux + 1 = 0$, so for each of the two values of u, we get a quadratic in x. $u = -1$ gives $x^2 + x + 1 = 0$, which has the roots $(-1 \pm \sqrt{3}i)/2$, while $u = -6/5$ gives $5x^2 + 6x + 5 = 0$, which has the roots $(-3 \pm 4i)/5$, so that our final solution set to the equation is

$$x \in \left\{ \frac{-1 + \sqrt{3}i}{2}, \frac{-1 - \sqrt{3}i}{2}, \frac{-3 + 4i}{5}, \frac{-3 - 4i}{5} \right\}.$$

Remark:

The solutions to this equation were deliberately chosen to be obscure, so as to thwart any attempt to factorize by inspection.

6. A calendar month is said to be "bad" if it has five Mondays. What is the maximum and minimum number of "bad" months in a year?

Solution:

If we assume that the year starts on day one, then we can determine all the months which have five of any given day. The results are summarised in the table below, where a month is listed under a day if and only if there are five of that day in the month:

	Day 1	Day 2	Day 3	Day 4	Day 5	Day 6	Day 7
Normal Year	Jan Apr July Oct Dec	Jan May July Oct	Jan May Aug Oct	Mar May Aug Nov	Mar June Aug Nov	Mar June Sept Dec	Apr July Sept Dec

	Day 1	Day 2	Day 3	Day 4	Day 5	Day 6	Day 7
Leap Year	Jan Apr July Sept Dec	Jan Apr July Oct Dec	Jan May July Oct	Feb May Aug Oct	Mar May Aug Nov	Mar June Aug Nov	Mar June Sept Dec

So, if the year starts on a Monday (or if it is a leap year and starts on a Sunday) then there will be five "bad" months in the year, and otherwise there will be four.

Obviously, it is easy to confirm that there exist both years which start with Mondays and years which do not, so that these upper and lower bounds are realized as the maximum and minimum respectively.

7. Evaluate
$$\sum_{r=1}^{n} r^4.$$

Solution:

We can easily get a telescoping form for a sum that will include r^4 by considering $(r+1)^5 - r^5 = 5r^4 + 10r^3 + 10r^2 + 5r + 1$. Summing this from $r = 1$ up to n, we get

$$(n+1)^5 - 1 = \sum_{r=1}^{n} \left[(n+1)^5 - n^5\right]$$
$$= 5\sum_{r=1}^{n} r^4 + 10\sum_{r=1}^{n} r^3 + 10\sum_{r=1}^{n} r^2 + 5\sum_{r=1}^{n} r + \sum_{r=1}^{n} 1$$

But now, we know that

$$\sum_{r=1}^{n} r^3 = \frac{n^2(n+1)^2}{4}, \quad \sum_{r=1}^{n} r^2 = \frac{n(n+1)(2n+1)}{6},$$

$$\sum_{r=1}^{n} r = \frac{n(n+1)}{2} \text{ and } \sum_{r=1}^{n} 1 = n.$$

Inserting all of these and solving for $\sum_{r=1}^{n} r^4$, we get

$$\sum_{r=1}^{n} r^4 = \frac{n(n+1)(2n+1)(3n^2+3n-1)}{30}.$$

Alternative Solution:

In order to sum over an arbitrary range, we would like to write r^4 in the form $f(r) - f(r-1)$ (which would give a telescoping sum). Now, by the Mean Value Theorem, we know that $f(r) - f(r-1) = f'(c)$ for some $c \in (r-1, r)$. Based on this, we might take $f_1(r) = r^5/5$ as a good approximation (since $f_1'(r) = r^4$). This gives us

$$f_1(r) - f_1(r-1) = r^4 - 2r^3 + 2r^2 - r + \frac{1}{5}.$$

Now, to improve the approximation, we would like to remove the $-2r^3$ term from the difference. Motivated again by the Mean Value Theorem, we might take $f_2(r) = r^5/5 - r^4/2$ as an improved approximation, since $f_2'(r) = r^4 - 2r^3$. This refinement gives us

$$f_2(r) - f_2(r-1) = r^4 - r^2 + r - \frac{3}{10}.$$

Continuing in the same manner, we set $f_3(r) = r^5/5 + r^4/2 + r^3/3$, which gives $f_3(r) - f_3(r-1) = r^4 + 1/30$. Finally then, we set

$$f(r) = \frac{r^5}{5} + \frac{r^4}{2} + \frac{r^3}{3} - \frac{r}{30}$$

to get $f(r) - f(r-1) = r^4$.

Substituting this expression into the sum, we get

$$\sum_{r=1}^{n} r^4 = \sum_{r=1}^{n} (f(r) - f(r-1)) = f(n) - f(0) = \frac{n^5}{5} + \frac{n^4}{2} + \frac{n^3}{3} - \frac{n}{30}.$$

and this is equal to the polynomial in the first solution.

8. Prove that the distance of the point whose coordinates are (x_1, y_1) from the line whose equation is $ax + by + c = 0$ is

$$\left| \frac{ax_1 + by_1 + c}{\sqrt{a^2 + b^2}} \right|,$$

Solution 1:
If the given line is vertical or horizontal, the result is simple to confirm — so we assume that a and b are both non-zero. Let $A = (x_1, y_1)$, and let $B = (x_2, y_2)$ be the point on the given line which is closest to A. Let $C = (x_3, y_1)$ be the intersection of the given line and a horizontal line passing through A. Then $x_3 = -(by_1 + c)/a$, and $|AC| = |(ax_1 + by_1 + c)/a|$.

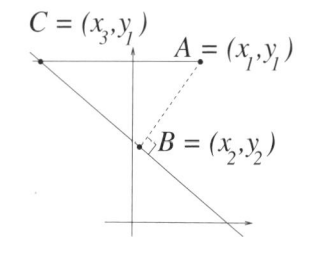

But now, since B is the closest point on the line to A, $\triangle ABC$ must be right-angled at B. So $|AB|/|AC| = \sin\theta$, where $\theta = \tan^{-1}(-a/b)$ is the angle of inclination of the given line.

But $\sin(\tan^{-1} x) = x/\sqrt{x^2+1}$, so $|\sin\theta| = |a/\sqrt{a^2+b^2}|$. Substituting this into the above equation, we get that the distance of the point (x_1, y_1) from the line $ax + by + c = 0$ is

$$|AB| = |AC||\sin\theta| = \left|\frac{ax_1 + by_1 + c}{a}\right| \left|\frac{a}{\sqrt{a^2+b^2}}\right| = \left|\frac{ax_1 + by_1 + c}{\sqrt{a^2+b^2}}\right|.$$

Solution 2:

A neat solution can be achieved using the close relationship between complex numbers and points in the plane. If (p_j, q_j) are points, and $r_j = p_j + iq_j$ for $j = 1, 2$ then the distance between the two points is simply $|r_2 - r_1|$.

If we choose a point (x, y) on the given line and let $z = x + iy$, $z_1 = x_1 + iy_1$, then we have

$$|ax_1 + by_1 + c| = |a(x_1 - x) + b(y_1 - y)| = \left|\operatorname{Re}(a + ib)\overline{(z_1 - z)}\right|. \quad (1)$$

But now, for any complex number z, $|\operatorname{Re} z| \leq |z|$, with equality if and only if $\operatorname{Im} z = 0$. So, we have

$$|ax_1 + by_1 + c| \leq |a + ib||z_1 - z|.$$

But $|a + ib| = \sqrt{a^2 + b^2}$, and $|z_1 - z|$ is the distance from (x_1, y_1) to a point on the line. Rearranging, we get that this distance must be at least

$$\left|\frac{ax_1 + by_1 + c}{\sqrt{a^2+b^2}}\right|,$$

with equality if and only if $b(x_1 - x) = a(y_1 - y)$ (so, the point on the line $ax + by + c = 0$ nearest to (x_1, y_1) is the intersection of this line with the line perpendicular to it passing through (x_1, y_1)).

Solution 3:

If the solver is familiar with the Cauchy-Schwarz inequality, this can be applied directly to the second term in (1) above to get

$$|ax_1 + by_1 + c| = |a(x_1 - x) + b(y_1 - y)| \leq \sqrt{a^2+b^2}\sqrt{(x_1-x)^2 + (y_1-y)^2},$$

which gives the result as in solution 2, and also provides the same equality condition.

9. Evaluate
$$\int_0^1 \sqrt{x - x^2}\,dx.$$

Solution:

To simplify this integral, we complete the square in the quadratic under the square root. This gives

$$\sqrt{x - x^2} = \sqrt{\frac{1}{4} - \left(x - \frac{1}{2}\right)^2} = \frac{\sqrt{1 - (2x - 1)^2}}{2}.$$

Substituting this into the integral, and making the substitution $u = 2x - 1$, we get

$$\int_0^1 \sqrt{x - x^2}\,dx = \frac{1}{2}\int_0^1 \sqrt{1 - (2x - 1)^2}\,dx = \frac{1}{4}\int_{-1}^1 \sqrt{1 - u^2}\,du.$$

There are a number of ways to evaluate the last integral. The standard approach is to make the substitution $u = \sin v$, which gives

$$\int_{-1}^1 \sqrt{1 - u^2}\,du = \int_{-\frac{\pi}{2}}^{\frac{\pi}{2}} \cos^2 v\,dv = \frac{1}{2}\int_{-\frac{\pi}{2}}^{\frac{\pi}{2}} (1 + \cos 2v)\,dv = \frac{\pi}{2}.$$

A geometric interpretation gives the same result in a neater manner: we need only observe that the integral is simply the area between the x-axis and the part of the unit circle in the upper half plane — so its value must simply be half the area of the unit circle, or $\frac{\pi}{2}$.

In any case, substituting the value of this integral into the above equation gives the value of the original integral:

$$\int_0^1 \sqrt{x - x^2}\,dx = \frac{1}{4}\int_{-1}^1 \sqrt{1 - u^2}\,du = \frac{\pi}{8}.$$

Alternative Solution:

We could also go straight to a trigonometric substitution, and set $x = \sin^2 u$. This gives $\sqrt{x} = \sin u$, $\sqrt{1 - x} = \cos u$ and $dx = 2\cos u \sin u\,du$. So, we get

$$\int_0^1 \sqrt{x - x^2}\,dx = \int_0^{\frac{\pi}{2}} 2\cos^2 u \sin^2 u\,du = \frac{1}{2}\int_0^{\frac{\pi}{2}} \sin^2 2u\,du.$$

Now, using the half-angle identity $1 - \cos 2\theta = 2\sin^2 \theta$, we get

$$\int_0^1 \sqrt{x-x^2}\,dx = \frac{1}{2}\int_0^{\frac{\pi}{2}} \sin^2 2u \, du = \frac{1}{4}\int_0^{\frac{\pi}{2}} (1-\cos 4u)\, du = \frac{\pi}{8}$$

as before.

10. If H is the orthocentre of a triangle ABC and O is the circumcentre, prove that
$$\overrightarrow{OA} + \overrightarrow{OB} + \overrightarrow{OC} = \overrightarrow{OH}$$
and
$$\overrightarrow{HA} + \overrightarrow{HB} + \overrightarrow{HC} = 2\overrightarrow{HO}.$$

Solution:

Let G be the centroid of the triangle. It is well known that $\vec{G} = \frac{\vec{A}+\vec{B}+\vec{C}}{3}$. Moreover, by Euler's famous result, O, G and H are collinear, with G lying between O and H, and with $|GH| = 2|OG|$. In vector form, this means that $2\vec{O} + \vec{H} = 3\vec{G}$.

Substituting for G, we get $2\vec{O} + \vec{H} = \vec{A} + \vec{B} + \vec{C}$. Subtracting $3\vec{O}$ from both sides, and using the identity $\vec{Y} - \vec{X} = \overrightarrow{XY}$, this gives
$$\overrightarrow{OH} = \overrightarrow{OA} + \overrightarrow{OB} + \overrightarrow{OC}.$$

Subtracting $3\vec{H}$ from both sides of the same identity gives
$$2\overrightarrow{HO} = \overrightarrow{HA} + \overrightarrow{HB} + \overrightarrow{HC}.$$

Eighth Superbrain, 1991 — Solutions

1. Show that no term of the sequence

$$11, 111, 1111, 11111, \ldots$$

   is the square of an integer.

   Solution:

   We are asked to show that certain numbers are not squares. One easy way to do this is using modular arithmetic. Now, if $k|100$ for some positive integer k, then all of the numbers in the sequence will have the same value modulo k (since they all end in the same digits, and hence differ by multiples of 100).

   But now, we know that $m^2 \equiv 0$ or $1 \bmod 4$ for any $m \in \mathbb{N}$, whereas $11 \equiv 3 \bmod 4$. So, since four is a factor of one hundred, we have $100n + 11 = 4(25k + 2) + 3 \equiv 3 \bmod 4$, and hence no term of the sequence is a square.

2. Find the 20th derivative, $\frac{d^{20}y}{dx^{20}}$, of the function

$$y = \frac{1}{1 - x^2}.$$

   Solution:

   To simplify the calculations, we may write y as a sum of partial fractions. Obviously, the denominator factors as a difference of two squares, so that we can write

$$y = \frac{1}{1 - x^2} = \frac{A}{x + 1} + \frac{B}{x - 1}.$$

   Multiplying out and solving, we get $A = -B = \frac{1}{2}$. So, since differentiation is a linear operator, we get

$$\frac{d^{20}y}{dx^{20}} = \frac{1}{2}\frac{d^{20}}{dx^{20}}\left(\frac{1}{x+1}\right) - \frac{1}{2}\frac{d^{20}}{dx^{20}}\left(\frac{1}{x-1}\right).$$

   But now, a simple induction establishes that

$$\frac{d^n}{dx^n}\left(\frac{1}{x+a}\right) = (-1)^n \frac{n!}{(x+a)^{n+1}},$$

   so that

$$\frac{d^{20}y}{dx^{20}} = \frac{20!}{2}\left(\frac{1}{(x+1)^{21}} - \frac{1}{(x-1)^{21}}\right).$$

3. The lengths of the sides of a given quadrilateral are a, b, c and d (labelled clockwise) and its area is A. Prove that $4A \leq (a+c)(b+d)$, with equality if and only if the figure is a rectangle.

Solution:

We label the vertices and angles of the quadrilateral as follows:

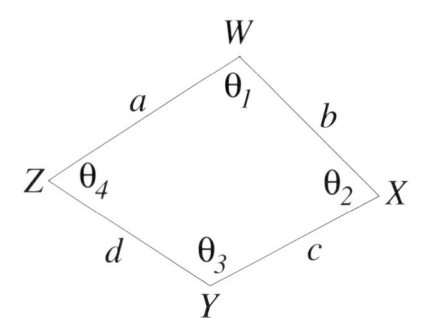

Let $\Delta(TUV)$ be the area of the triangle with vertices at T, U and V. Then $2A = 2\Delta(WXY) + 2\Delta(WZY) = bc\sin\theta_2 + ad\sin\theta_4 \leq bc + ad$, with equality if and only if $\sin\theta_2 = \sin\theta_4 = 1$. Similarly, $2A \leq ab + cd$, with equality if and only if $\sin\theta_1 = \sin\theta_3 = 1$.

So, adding the two inequalities, we get

$$4A \leq ab + cd + bc + ad = (a+c)(b+d),$$

with equality if and only if $\sin\theta_i = 1$ for each i. But now, $\sin\theta_i = 1$ if and only if θ_i is a right angle, so the only case of equality is when each angle is a right angle, and hence the quadrilateral is a rectangle.

Remark:

Although we do not explicitly consider the case of a non-convex quadrilateral here, note that the identities and the inequalities used in the solution all still hold if $\theta_i > \pi$ for some i.

4. Evaluate
$$\lim_{n\to\infty} \left[\sqrt{(n+4)(n+5)} - \sqrt{(n+2)(n+3)}\right].$$

Solution:

Any time we see square roots in a limit, the natural response is to try to remove these. One way to do this is to multiply above and below by its conjugate. In this case, we multiply above and below by

$\sqrt{(n+4)(n+5)} + \sqrt{(n+2)(n+3)}$, and the function whose limit we are seeking simplifies to

$$\frac{4n+14}{\sqrt{(n+4)(n+5)} + \sqrt{(n+2)(n+3)}}.$$

Dividing above and below by n, we get

$$\frac{4 + \frac{14}{n}}{\sqrt{(1+\frac{4}{n})(1+\frac{5}{n})} + \sqrt{(1+\frac{2}{n})(1+\frac{3}{n})}}.$$

Now, each of the separate parts has a non-zero limit, so we can combine each of the separate limits to get

$$\lim_{n\to\infty}\left[\sqrt{(n+4)(n+5)} - \sqrt{(n+2)(n+3)}\right] = \frac{4}{\sqrt{(1)(1)} + \sqrt{(1)(1)}} = 2.$$

Alternative Solution:

Let $f(x) = \sqrt{x(x+1)}$. Then by the Mean Value Theorem, there exists $c \in (n+2, n+4)$ such that

$$\sqrt{(n+4)(n+5)} - \sqrt{(n+2)(n+3)} = 2f'(c) = \frac{2c+1}{\sqrt{c^2+c}} = \frac{2+\frac{1}{c}}{\sqrt{1+\frac{1}{c}}}.$$

But now, $c > n+2$, so as n grows without bound, c must also grow without bound, so that

$$\lim_{n\to\infty}\left(\sqrt{(n+4)(n+5)} - \sqrt{(n+2)(n+3)}\right) = \lim_{c\to\infty}\frac{2+\frac{1}{c}}{\sqrt{1+\frac{1}{c}}} = 2.$$

5. Find the nth term of the sequence

$$\sum_{r=1}^{\infty}\frac{1}{r(r+1)}, \quad \sum_{r=1}^{\infty}\frac{1}{r(r+1)(r+2)}, \quad \sum_{r=1}^{\infty}\frac{1}{r(r+1)(r+2)(r+3)} \cdots$$

Solution:

Let $T_n = \sum_{r=1}^{\infty}\frac{1}{\prod_{k=0}^{n}(r+k)}$, and $T_{n,m} = \sum_{r=1}^{m}\frac{1}{\prod_{k=0}^{n}(r+k)}$.

We can investigate any one of the terms by writing the summand in partial fractions to get a telescoping sum. For instance,

$$\sum_{r=1}^{\infty} \frac{1}{r(r+1)} = \sum_{r=1}^{\infty} \left(\frac{1}{r} - \frac{1}{r+1}\right) = \lim_{k \to \infty} \left(1 - \frac{1}{k+1}\right) = 1.$$

Evaluating the first few terms gives $T_1 = 1$, $T_2 = \frac{1}{4}$, $T_3 = \frac{1}{18}$, $T_4 = \frac{1}{96}$, ...
This leads us to the conjecture that $T_n = 1/(n.n!)$. To prove this conjecture, we investigate the partial sums. Again, by writing the summand as partial fractions, we may evaluate the partial sums explicitly. This gives

$$\frac{1}{1.1!} - T_{1,m} = \frac{1}{m+1}, \quad \frac{1}{2.2!} - T_{2,m} = \frac{1}{2(m+1)(m+2)},$$

$$\frac{1}{3.3!} - T_{3,m} = \frac{1}{3(m+1)(m+2)(m+3)}, \ldots$$

So, we are lead to the conjecture that

$$T_{n,m} = \frac{1}{n.n!} - \frac{m!}{n(m+n)!}.$$

We prove this by induction on m.

For $m = 1$,
$$T_{n,1} = \frac{1}{(n+1)!} = \frac{1}{n.n!} - \frac{1}{n.(n+1)!},$$
so the conjecture is true. Now, assume that the conjecture is true for $m = k-1$ — that is, that

$$T_{n,k-1} = \frac{1}{n.n!} - \frac{(k-1)!}{n(k+n-1)!}.$$

Then
$$T_{n,k} = \frac{1}{n.n!} - \frac{(k-1)!}{n(k+n-1)!} + \frac{(k-1)!}{(k+n)!} = \frac{1}{n.n!} - \frac{k!}{n(n+k)!},$$

so the conjecture is true for $m = k$, and hence by induction for all positive integers m.

Now that we have a closed form for the partial sums, we need only take the limit as n tends to infinity:

$$T_n = \lim_{m \to \infty} T_{n,m} = \frac{1}{n.n!} - \lim_{m \to \infty} \frac{m!}{n(m+n)!} = \frac{1}{n.n!}.$$

6. Solve the equations

(a) $\dfrac{2x+1}{(x+3)(x-2)} + \dfrac{5x}{(2x+1)(x+3)} = \dfrac{x+3}{(2x+1)(x-2)}.$

(b) $3\left(\sin^4\theta + \cos^4\theta\right) - 2\left(\sin^6\theta + \cos^6\theta\right) = 1.$

Solution:

(a) For all the terms to be defined, we must have $x \notin \{-3, -1/2, 2\}$. So, we may gather the terms to one side and multiply through by $(x-2)(x+3)(2x+1)$ to get
$$(2x+1)^2 + 5x(x-2) - (x+3)^2 = 8x^2 - 12x - 8 = 0.$$
However, this quadratic has only two real roots, $x = -1/2$ or $x = 2$, and since we have already established that neither of these is a solution, there are no solutions to the equation given.

(b) This problem is complicated by having terms both of degree four and degree six on the left-hand side. To remedy this, we multiply the fourth power terms by $1 = \sin^2\theta + \cos^2\theta$ (since this makes no change to the value). Writing $s = \sin\theta$, $c = \cos\theta$ to simplify the calculations, we get
$$3(s^4+c^4)(s^2+c^2) - 2(s^6+c^6) = s^6 + 3s^4c^2 + 3s^2c^4 + c^6 = (s^2+c^2)^3.$$
But again, $s^2+c^2 = 1$, so $3\left(\sin^4\theta + \cos^4\theta\right) - 2\left(\sin^6\theta + \cos^6\theta\right) = 1$ must hold for all θ.

Note that we could also remedy the problem the different degrees by writing $s^6 + c^6 = (s^2+c^2)(s^4 - s^2c^2 + c^4) = s^4 - s^2c^2 + c^4$, from which the equation to be solved reduces to $s^4 + 2s^2c^2 + c^4 = (s^2+c^2)^2 = 1$, which is also an identity.

7. If $x+y+z = 0$, prove that
$$7(x^2+y^2+z^2)(x^5+y^5+z^5) = 10(x^7+y^7+z^7).$$

Solution:

Let $z = -x - y$. Then $x^2+y^2+z^2 = 2(x^2+xy+y^2)$, $x^5 + y^5 + z^5 = -5(x^4y + 2x^3y^2 + 2x^2y^3 + xy^4)$, $x^7+y^7+z^7 = -7(x^6y + 3x^5y^2 + 5x^4y^3 + 5x^3y^4 + 3x^2y^5 + xy^6)$, and $7(x^2+y^2+z^2)(x^5+y^5+z^5)$
$= -70(x^2+xy+y^2)(x^4y + 2x^3y^2 + 2x^2y^3 + xy^4)$
$= -70(x^6y + 3x^5y^2 + 5x^4y^3 + 5x^3y^4 + 3x^2y^5 + xy^6)$
$= 10(x^7+y^7+z^7).$

Alternative Solution:

Another way to deal with the constraint $x+y+z=0$ is to introduce a substitution which deals with it automatically in a neat way. One such substitution is

$$x = a+b, \quad y = a\omega + b\omega^2, \quad z = a\omega^2 + b\omega,$$

where ω is a complex cube root of unity.

With this substitution, the algebra simplifies considerably. We get

$$x^n + y^n + z^n = \sum_{r=0}^{n} \binom{n}{r} a^r b^{n-r} \left(\sum_{\alpha=0}^{2} (\omega^{(2n-r)})^\alpha \right).$$

Now, it is easy to see that the inner sum will be equal to $1+\omega+\omega^2 = 0$ unless $2n-r$ is a multiple of 3, in which case it will be equal to 3. So we get

$$x^n + y^n + z^n = 3 \sum \binom{n}{r} a^r b^{n-r},$$

where the sum is taken over all integers $0 \le r \le n$ such that $3 | (2n-r)$. Setting $n = 2, 5$ and 7, we get:

$$x^2 + y^2 + z^2 = 6ab,$$

$$x^5 + y^5 + z^5 = 15(ab^4 + a^4 b),$$

$$x^7 + y^7 + z^7 = 63(a^2 b^5 + a^5 b^2),$$

and the result follows easily.

It is fair to ask if the substitution was conjured out of thin air. Well, since $1 + \omega + \omega^2 = 0$, the condition $x + y + z = 0$ follows immediately. It is easy to confirm that such a pair (a, b) exists for each triple (x, y, z) with $x + y + z = 0$. In fact, it is not hard to construct them explicitly by solving simultaneous equations; a and b can be expressed in terms of x and y as follows:

$$a = \frac{x\omega^2 - y}{\omega(\omega - 1)}, \quad b = \frac{x\omega - y}{\omega(1 - \omega)}.$$

8. Let m, n and k be positive integers. One solution of the equation $(m!)(n!) = k!$ is $m = 6$, $n = 7$, $k = 10$. Find another solution with m, n and k all greater than 10.

Solution:

For any positive integer n, we have $n! = n(n-1)!$, so that we have $a = bc$, with two of a, b and c being factorials. To complete the solution, we need only let $n = r!$ for some positive integer r, giving

$$(r!)! = r!(r!-1)!.$$

So, for any $r > 10$, let $\{m, n, k\} = \{r, r!-1, r!\}$ and we have an infinite set of solutions.

Remark 1:

This is a classic diversion technique — the example given in the question is quite misleading, since it is very difficult if not impossible to find more solutions like it, especially for larger numbers. The question is, however, quite easily solved by taking a step back.

Remark 2:

It is not known whether there are any ordered triples (m, n, k) with $(m!)(n!) = k!$ which are not of the form $(r, r!-1, r!)$ for some positive integer r other than the example given in the question.

9. If $f(x) = x/2 + \cos x$ and $x \in [0, 2\pi]$, what are the greatest and least values of $f(x)$?

Solution:

Since f is continuous and differentiable everywhere in its domain of definition, its maximum and minimum values must occur at the extremes of its range, or at the zeros of its derivative.

$f'(x) = \frac{1}{2} - \sin x$, so $f'(x) = 0$ if and only if $x = \pi/6$ or $x = 5\pi/6$ (since $x \in [0, 2\pi]$). So the maximum and minimum values of f must be in the set

$$f\left(\left\{0, \frac{\pi}{6}, \frac{5\pi}{6}, 2\pi\right\}\right) = \left\{1, \frac{\pi}{12} + \frac{\sqrt{3}}{2}, \frac{5\pi}{12} - \frac{\sqrt{3}}{2}, 1+\pi\right\}.$$

A simple inspection shows that

$$f_{min} = \frac{5\pi}{12} - \frac{\sqrt{3}}{2} \text{ and } f_{max} = 1 + \pi.$$

10. Prove that $4^{50} + 2^{50} + 1$ is not a prime number.

Solution:

The easiest way to show that a number is not prime is to find a non-trivial factor. The large exponents need not be of too much concern, since we wish to evaluate $4^{50}+2^{50}+1$ modulo p for some prime number, and exponents can be easily handled with modular arithmetic.

A brief inspection shows that $p = 3$ provides the solution that we need, since
$$4^{50} + 2^{50} + 1 \equiv 1^{50} + (-1)^{50} + 1 \equiv 0 \bmod 3.$$

So $3 \mid (4^{50} + 2^{50} + 1)$, and clearly $3 < 4^{50} + 2^{50} + 1$, so that $4^{50} + 2^{50} + 1$ cannot be prime.

Alternative Solution:

In the previous solution, we found a specific factor. However, it is also possible to factorize $4^{50} + 2^{50} + 1$ more generally, as an expression of the form $x^{4n} + x^{2n} + 1$. Completing the square in x^{2n}, we get

$$x^{4n} + x^{2n} + 1 = \left(x^{4n} + 2x^{2n} + 1\right) - x^{2n}.$$

But now, x^{2n} is itself a square, so we can factorize the difference of two squares to get

$$x^{4n} + x^{2n} + 1 = \left(x^{2n} + x^n + 1\right)\left(x^{2n} - x^n + 1\right).$$

Finally, it is easy to see that in this particular case, both of $2^{50} + 2^{25} + 1$ and $2^{50} - 2^{25} + 1$ are non-trivial factors, and so their product is clearly not prime.

Remark:

In general, $x^{2n} + x^n + 1 > x^{2n} - x^n + 1 > 1$ for $x > 1$, so the only prime of the form $x^{4n} + x^{2n} + 1$ is 3.

Ninth Superbrain, 1992 — Solutions

1. What, in your opinion, is the next term in each of the following sequences and why?

   (a) T, Q, P, H, H, O, N?

   (b) 2, 3, 4, 5, 7, 8, 9, 11, 13, 17?

   (c) 1, 2, 6, 2, 1, 7, 5, 4, 3, 3, 3, 4, 6?

   (d) 1827, 6412, 5216, 3435, 1272, 9100?

   (e) 1, 3, 11, 25, 137, 49, 363, 761?

   Solution:

   (a) This sequence is the first letter of the word for an n-sided figure for $n = 3, 4, 5, \ldots$ — i.e. Triangle, Quadrilateral, Pentagon, Hexagon, Heptagon, Octagon, Nonagon.

   The next figure is a ten-sided figure, or a Decagon, so the missing next term of the sequence is 'D'.

   (b) If we consider the prime factorisation of the terms of this sequence, we see that each one contains only a single prime. Conversely, we note that all prime powers up to 17 are included, and we conclude that this is the sequence of powers of prime numbers. The next term is therefore 19.

   (c) The first three terms in this sequence are the values of $n!$ for $n = 1, 2, 3$, which may prompt the observant solver. Based on this, we would expect the next few terms to be 24, 120, 720, whereas instead they are 2, 1, 7 — but having come this far, we notice that the terms which do appear are each the first digit of the corresponding factorial. It is easy to confirm that this hypothesis matches the remaining terms, and therefore that the missing next term must be 8, the first digit of $14! = 87178291200$.

   (d) Although each term in this sequence is a four-digit number, the key to solving it is to ignore the groupings as they are written. In fact, it may be easier to see the pattern with each of the digits evenly spaced:

   $$1\ 8\ 2\ 7\ 6\ 4\ 1\ 2\ 5\ 2\ 1\ 6\ 3\ 4\ 3\ 5\ 1\ 2\ 7\ 2\ 9\ 1\ 0\ 0$$

   The observant solver (perhaps after trying a number of familiar sequences) will notice that the first few digits are the cubes of the first few numbers: 1, 8, 27, 64,...

So we are lead to a hypothesis that the sequence is the sequence of positive integer cubes, grouped in blocks of four digits. It is easy to confirm that this hypothesis is borne out by the remaining terms, and the missing next term is simply the last digit of $10^3 = 1000$ and the first three digits of $11^3 = 1331$ — i.e. 0133.

(e) Without either a familiarity with the underlying sequence or a moment of inspiration, it is hard to see how any solver would crack this pattern in the exam (indeed, that may have been precisely the reason that it was set only as the fifth part of the question).

The underlying is that of harmonic sums, $H_n = \sum_{k=1}^{n} \frac{1}{k}$, which begins

$$1, \quad \frac{3}{2}, \quad \frac{11}{6}, \quad \frac{25}{12}, \quad \frac{137}{60}, \quad \frac{49}{20}, \quad \frac{363}{140}, \quad \frac{761}{280}, \quad \frac{7129}{2520},$$

and the sequence in question is the sequence of numerators of H_n (written in fully reduced form). The missing next term is 7129.

2. Let $p_1 < p_2 < p_3$ be prime numbers such that $p_1^2 + p_2^2 + p_3^2$ is also a prime number. Find the value of p_1.

Solution:

Let $p = p_1^2 + p_2^2 + p_3^2$.

Obviously p_2 and p_3 are odd, since neither of them can be two. But then, if $p_1 = 2$, p would be even, and hence not prime — so $p_1 \geq 3$. This means that $p_2, p_3 > 3$. But any prime $q > 3$ satisfies $q \equiv \pm 1 \bmod 6$, and hence $q^2 \equiv 1 \bmod 6$. So, if $p_1 > 3$, then $p \equiv 3 \bmod 6$, and hence $3 | p$. But obviously $p > 3$, so in this case p cannot be prime.

This leaves only the case where $p_1 = 3$, which must be the solution.

We observe that $p_1 = 3$ is a possible solution since, for example, $\{p_1, p_2, p_3\} = \{3, 5, 7\}$ gives $p = 83$, which is indeed prime.

3. Find any triangle ABC such that each side has an integer length, and one angle of the triangle is twice the size of another angle of the triangle.

Solution:

Let the side lengths of the triangle be a, b and c, and the angles opposite these sides be α, β and γ respectively. We may assume that $\beta = 2\alpha$.

By the sine rule,

$$\frac{a}{\sin \alpha} = \frac{b}{\sin \beta} = \frac{b}{2 \sin \alpha \cos \alpha},$$

and so we require that $\cos \alpha = b/2a$.

By the cosine rule,
$$\frac{b}{2a} = \cos \alpha = \frac{b^2 + c^2 - a^2}{2bc},$$

and rearranging, it suffices that $b^2 = a(c + a)$ providing that $c \neq a$ (a division by $c - a$ is involved).

To finish the problem, all we need is a single solution to the equation which satisfies the triangle inequality. A simple investigation shows that
$$(a, b, c) = (4, 6, 5)$$
is a small solution.

Remark 1:

Although it is not necessary to solve the question, it is possible to find the general solution.

Let $d = \gcd(a, c)$, and let $a = a'd$, $b = b'd$, $c = c'd$. Then $b'^2 = a'(a'+c')$. Since a and $a + c$ are relatively prime, we must have $a' = n^2$ for some positive integer n, $a' + c' = m^2$ for some integer $m > n$, and $b' = mn$. Let $k = m - n$, and we get the general solution to the equation:
$$(a, b, c) = (n^2 d, (n + k)nd, (k^2 + 2kn)d),$$
where d, n and k are any positive integers.

If we additionally want a, b and c to be the sides of a triangle, then they must satisfy the triangle inequality. (Since d is common to all three, we may assume that $d = 1$.)

$b + c > a$ is equivalent to $k^2 + 3kn + n^2 > n^2$, which follows immediately from $n, k > 0$.

$a + c > b$ is equivalent to $k^2 + 2kn + n^2 > n^2 + kn$, which also follows immediately from $n, k > 0$.

$a + b > c$ is equivalent to $2n^2 > k^2 + kn$, or $\left(\frac{3n}{2}\right)^2 > \left(k + \frac{n}{2}\right)^2$. Since both terms are positive, we may take square roots, to get $\frac{3n}{2} > k + \frac{n}{2}$, or $n > k$.

So, all triangles with integer side lengths and one angle equal to twice another angle have the form
$$(a, b, c) = (n^2 d, (n + k)nd, (k^2 + 2kn)d),$$

where d and n are arbitrary positive integers, and k is a positive integer less than n.

Remark 2:

Another way to derive the general form of a, b and c is to observe that $b^2 = a(a+c)$ is equivalent to $(2b)^2 + c^2 = (2a+c)^2$. So, $(2b, c, 2a+c)$ is a Pythagorean triple and use the well-known formula for generating these. However, this requires considering two cases (since $2b$ and c can be interchanged) and also separately imposing the condition that a, b and c be the side-lengths of a triangle in order to arrive back at the form derived above.

4. Evaluate
$$\sum_{r=1}^{\infty} \tan^{-1}\left(\frac{1}{2r^2}\right).$$

Solution:

In order to add inverse tangents, we need to manipulate the tan addition rule (unless the solver is already familiar with the addition rule for inverse tangents):

$$\tan(a+b) = \frac{\tan a + \tan b}{1 - \tan a \tan b}.$$

Letting $a = \tan^{-1} \alpha$ and $b = \tan^{-1} \beta$ and taking the inverse tangents of both sides, we get

$$\tan^{-1} \alpha + \tan^{-1} \beta = \tan^{-1}\left(\frac{\alpha + \beta}{1 - \alpha\beta}\right).$$

Let $T_k = \sum_{r=1}^{k} \tan^{-1} \frac{1}{2r^2}$. Evaluating the first handful of terms gives $T_1 = \tan^{-1} \frac{1}{2}$, $T_2 = \tan^{-1} \frac{2}{3}$, $T_3 = \tan^{-1} \frac{3}{4}$, leading to the conjecture $T_k = \tan^{-1} \frac{k}{k+1}$. We prove this conjecture by induction.

For $k = 1$, the claim is certainly true. Now assume that it is true for $k = m - 1$ — that is, that $T_{m-1} = \tan^{-1} \frac{m-1}{m}$. Then

$$\begin{aligned} T_m &= \tan^{-1} \frac{m-1}{m} + \tan^{-1} \frac{1}{2m^2} = \tan^{-1} \frac{\frac{m-1}{m} + \frac{1}{2m^2}}{1 - \left(\frac{m-1}{m}\right)\left(\frac{1}{2m^2}\right)} \\ &= \tan^{-1} \frac{m(2m^2 - m + 1)}{(m+1)(2m^2 - m + 1)} = \tan^{-1} \frac{m}{m+1}. \end{aligned}$$

So, by induction, $\sum_{r=1}^{k} \tan^{-1} \frac{1}{2k^2} = \tan^{-1} \frac{k}{k+1}$ for all positive integers k. So, since $\tan^{-1}$ is a continuous function, we may write

$$\sum_{r=1}^{\infty} \tan^{-1}\left(\frac{1}{2r^2}\right) = \lim_{k \to \infty} \tan^{-1}\left(\frac{k}{k+1}\right)$$
$$= \tan^{-1}\left(\lim_{k \to \infty} \frac{k}{k+1}\right) = \tan^{-1} 1 = \frac{\pi}{4}.$$

Alternative Solution:

An elegant and much more general proof is available to the reader who is familiar with some complex analysis. Let

$$f(x) = \sum_{n=1}^{\infty} \tan^{-1}\left(\frac{x}{n^2}\right).$$

Then

$$f'(x) = \sum_{n=1}^{\infty} \frac{n^2}{n^4 + x^2} = \frac{1}{2} \sum_{n=1}^{\infty} \left[\frac{1}{n^2 + ix} + \frac{1}{n^2 - ix}\right]$$
$$= \frac{1}{2i} \frac{d}{dx} \log\left[\prod_{n=1}^{\infty} \left(\frac{1 + \frac{ix}{n^2}}{1 - \frac{ix}{n^2}}\right)\right].$$

Using

$$\prod_{n=1}^{\infty} \left(1 - \frac{z^2}{n^2}\right) = \frac{\sin \pi z}{\pi z},$$

we get

$$f'(x) = \frac{1}{2i} \frac{d}{dx} \log\left[\frac{\sin e^{\frac{3i\pi}{4}} \sqrt{x\pi}}{i \sin e^{\frac{i\pi}{4}} \sqrt{x\pi}}\right].$$

Integrating with respect to x gives

$$f(x) = \text{Re}\left(\frac{1}{2i} \log\left[\frac{\sin e^{\frac{3i\pi}{4}} \sqrt{x\pi}}{i \sin e^{\frac{i\pi}{4}} \sqrt{x\pi}}\right] - c\right),$$

where $c =$

$$\lim_{x \to 0} \frac{1}{2i} \log\left[\frac{\sin e^{\frac{3i\pi}{4}} \sqrt{x\pi}}{i \sin e^{\frac{i\pi}{4}} \sqrt{x\pi}}\right] = \lim_{x \to 0} \frac{1}{2i} \log\left[\frac{e^{\frac{3i\pi}{4}} \sqrt{x\pi}}{ie^{\frac{i\pi}{4}} \sqrt{x\pi}}\right] = \lim_{x \to 0} \frac{1}{2i} \log(1) = 0.$$

So, we have

$$f(x) = f(x) - f(0) = \text{Re}\left(\frac{1}{2i}\log\left[\frac{\sin e^{\frac{3i\pi}{4}}\sqrt{x\pi}}{i\sin e^{\frac{i\pi}{4}}\sqrt{x\pi}}\right]\right),$$

and hence

$$\sum_{r=1}^{\infty}\tan^{-1}\left(\frac{1}{2r^2}\right) = f\left(\frac{1}{2}\right) = \text{Re}\left(-\frac{1}{2i}\log\left[\frac{\sin e^{\frac{3i\pi}{4}}\sqrt{\frac{1}{2}\pi}}{i\sin e^{\frac{i\pi}{4}}\sqrt{\frac{1}{2}\pi}}\right]\right)$$

$$= \frac{1}{2}\text{Im}\left(\log\left[\frac{\sin\left(-\frac{\pi}{2}+\frac{i\pi}{2}\right)}{i\sin\left(\frac{\pi}{2}+\frac{i\pi}{2}\right)}\right]\right) = \frac{1}{2}\text{Im}(\log(i)) = \frac{\pi}{4}.$$

5. Solve the equation

$$(x-1)(x-2)(x-3)(x-4) + 9 = 0.$$

Solution:

If we multiply $(x-1)$ by $(x-4)$ and $(x-2)$ by $(x-3)$, then both of the resulting quadratics will have the same coefficients of x^2 and of x (in particular, $(x-1)(x-4) = x^2 - 5x + 4$, $(x-2)(x-3) = x^2 - 5x + 6$). Let $u = x^2 - 5x + 5$. Then $(x-1)(x-4) = u-1$, $(x-2)(x-3) = u+1$, and the equation becomes $u^2 + 8 = 0$, which has the solutions $u = \pm 2\sqrt{2}i$. Putting this back into the definition of u, we get $x^2 - 5x + 5 \pm 2\sqrt{2}i = 0$, which we can solve using the quadratic formula to get the complete set of solutions for x:

$$\left\{\frac{5+\sqrt{5+8\sqrt{2}i}}{2}, \frac{5-\sqrt{5+8\sqrt{2}i}}{2}, \frac{5+\sqrt{5-8\sqrt{2}i}}{2}, \frac{5-\sqrt{5-8\sqrt{2}i}}{2}\right\}.$$

6. Find the maximum value of the function

$$f(\theta) = \sin^3\theta - 4\sin\theta.$$

Solution:

Since f is differentiable and periodic, it must attain its maximum value at a zero of its derivative. $f'(\theta) = (3\sin^2\theta - 4)\cos\theta$, so $f'(\theta) = 0$ if and only if $\cos\theta = 0$, or $\theta = n\pi + \frac{\pi}{2}$. Now $\sin\left(n\pi + \frac{\pi}{2}\right) = (-1)^n$, so $f\left(n\pi + \frac{\pi}{2}\right) = -3(-1)^n$, and hence $f_{max} = 3$.

Alternative Solution:

This problem can be approached in a slightly different manner by dispensing immediately with the sine function. Since the range of sine is $[-1, 1]$, the maximum value of $g(\sin\theta) = \sin^3\theta - 4\sin\theta$ is simply the maximum value of $g(x) = x^3 - 4x$ on the set $[-1, 1]$.

But now, $g'(x) = 3x^2 - 4 < 0$ for $|x| \leq 1$, so that the maximum value of $g(x)$ on the range must be at the lower bound of the range — i.e. $g_{min} = g(-1) = 3$.

7. Let P_i be distinct points with coordinates (x_i, y_i) for $i \in [1, n]$. Find the coordinates of a point P such that

$$\sum_{i=1}^{n} |PP_i|^2$$

is as small as possible.

Solution:

Let P have co-ordinates (x, y), and let

$$Q = \sum_{i=1}^{n} |PP_i|^2 = \sum_{i=1}^{n} \left[(x - x_i)^2 + (y - y_i)^2\right].$$

Completing the square in x and y, we get

$$\begin{aligned}
Q &= n\left(x - \tfrac{1}{n}\sum_i x_i\right)^2 + \sum_i x_i^2 - \tfrac{1}{n}\left(\sum_i x_i\right)^2 \\
&\quad + n\left(y - \tfrac{1}{n}\sum_i y_i\right)^2 + \sum_i y_i^2 - \tfrac{1}{n}\left(\sum_i y_i\right)^2 \\
&= n\left(x - \tfrac{1}{n}\sum_i x_i\right)^2 + n\left(y - \tfrac{1}{n}\sum_i y_i\right)^2 \\
&\quad + \sum_i \left(x_i - \tfrac{1}{n}\sum_j x_j\right)^2 + \sum_i \left(y_i - \tfrac{1}{n}\sum_j y_j\right)^2 \\
&\geq \sum_i \left(x_i - \tfrac{1}{n}\sum_j x_j\right)^2 + \sum_i \left(y_i - \tfrac{1}{n}\sum_j y_j\right)^2,
\end{aligned}$$

with equality if and only if $x = \tfrac{1}{n}\sum_i x_i$ and $y = \tfrac{1}{n}\sum_i y_i$. So, to minimize Q, we must have $P = \left(\dfrac{1}{n}\sum_{i=1}^{n} x_i, \dfrac{1}{n}\sum_{i=1}^{n} y_i\right)$.

Alternative Solution:

If the reader is familiar with partial differentiation, it is possible to analyse the minima of the function Q defined in the previous solution by that method. Moreover, note that Q clearly cannot be minimal for a point P outside a circle centred on the origin with radius $\max_{1 \leq i \leq n} \sqrt{x_i^2 + y_i^2}$, since each $|PP_i|$ could then be reduced by bringing P closer to the circle. So, to find the global maximum of Q, it is sufficient to find the maximum of Q on the (compact) set of points within this circle.

Fixing y and differentiating Q with respect to x gives

$$\frac{\partial Q}{\partial x} = 2\sum_i (x - x_i) = 2n\left(x - \frac{1}{n}\sum_i x_i\right).$$

But $\dfrac{\partial^2 Q}{\partial x^2} = 2n > 0$, so $x = \dfrac{1}{n}\sum_i x_i$ gives a minimum value for Q.

Similarly, $y = \dfrac{1}{n}\sum_i y_i$ gives the minimum with respect to y. So, to minimize Q, we must have $P = \left(\dfrac{1}{n}\sum_{i=1}^n x_i, \dfrac{1}{n}\sum_{i=1}^n y_i\right)$ as before.

Remark:

This question is a generalization of question 7 of 1985, although the same basic method still works.

8. Evaluate
$$\int \frac{dx}{\sin^4 x \cos^4 x}.$$

Solution:

We know that $\sin\theta \cos\theta = \frac{1}{2}\sin 2\theta$, so with the substitution $u = 2x$, we can write

$$\int \frac{dx}{\sin^4 x \cos^4 x} = \int \frac{16\, dx}{\sin^4 2x} = 8\int \frac{du}{\sin^4 u}.$$

Dividing above and below by $\cos^4 u$, we get

$$\frac{1}{\sin^4 u} = \frac{\sec^4 u}{\tan^4 u}.$$

But now, we know that $\frac{d\tan u}{du} = \sec^2 u$, and $\sec^2 u = 1 + \tan^2 u$

$$\int \frac{dx}{\sin^4 x \cos^4 x} = 8 \int \frac{1 + \tan^2 u}{\tan^4 u} \sec^2 u\, du = 8 \int \frac{1 + t^2}{t^4} dt,$$

where $t = \tan x$. But now, the last integral can easily be calculated, giving $\int \frac{1+t^2}{t^4} dt = -\frac{4}{t^3} - \frac{2}{t} = -\frac{4}{\tan^3 2x} - \frac{2}{\tan 2x}$. So we have

$$\int \frac{dx}{\sin^4 x \cos^4 x} = -\frac{4}{\tan^3 2x} - \frac{2}{\tan 2x} + C,$$

where C is an arbitrary constant.

9. If a, b, c and d are real numbers, prove that $a^4 + b^4 + c^4 + d^4 \geq 4abcd$, assuming only that $x^2 \geq 0$ for any real number x.

   Solution:

   $(a^2 - b^2)^2 \geq 0$, or $a^4 + b^4 \geq 2a^2b^2$. So $a^4 + b^4 + c^4 + d^4 \geq 2a^2b^2 + 2c^2d^2$. $(ab - cd)^2 \geq 0$, or $a^2b^2 + c^2d^2 \geq 2abcd$. So

   $$a^4 + b^4 + c^4 + d^4 \geq 2a^2b^2 + 2c^2d^2 \geq 4abcd.$$

10. If m and n are odd positive integers such that m^m divides n^n, prove or disprove that m divides n.

    Solution:

    The problem solver who is willing to bend the rules a little can gain a head start in this problem by realising that, from a point of view of divisibility, there is very little difference between 2 and any other prime. This suggests that the condition that the numbers be odd is an artificial restraint, perhaps to exclude a small solution.

    Indeed, a brief investigation reveals that there is a small even pair m, n such that m does not divide n, but m^m divides n^n — namely $m = 4$, $n = 10$.

    Studying this example carefully reveals a method which can then be used to generate an odd solution.

    We let $m = p^2$ for some prime number p, and let $n = pq$, where $q > 2p$ but $p \nmid q$. Then $m \nmid n$, since m is divisible by a higher power of p, but $m^m = p^{2p^2} \mid p^{pq}q^{pq} = n^n$, since $pq > 2p^2$.

    For instance, let $m = 9$, $n = 21$. Clearly m is not a factor of n, but $m^m = 3^{18}$ is a factor of $n^n = 3^{21}7^{21}$.

Tenth Superbrain, 1993 — Solutions

1. The length of each side of a rectangle is a whole number, and the area of the rectangle measured in square metres is equal to the length of the perimeter of the rectangle measured in metres. If the rectangle is not a square, find with proof the dimensions of the rectangle.

   Solution:

   Let the sides have lengths a metres and b metres respectively. Then the area of the rectangle is ab, and its perimeter is $2a + 2b$. So the condition becomes $ab = 2a + 2b$. Rearranging and factoring, we get
   $$ab - 2a - 2b + 4 = (a-2)(b-2) = 4.$$
   Now, four can be written as a product of two distinct integers in exactly one way (i.e. 4=1.4), so that we must have $\{a-2, b-2\} = \{1, 4\}$, or $\{a, b\} = \{3, 6\}$ (up to ordering). So the dimensions of the rectangle must be 3 metres × 6 metres.

2. Evaluate
   $$\sum_{r=1}^{n} (r^2 + 1) r!.$$

   Solution:

   If we evaluate the first few terms of the sequence, we get 2, 12, 72, 480, 3600. Motivated by the presence of the factorial in the problem, we observe that the sum of the first n terms is divisible by $(n+1)!$ but not by $(n+2)!$ for each of the first few terms.

   Moreover, if we evaluate the quotient
   $$\frac{\sum_{r=1}^{n}(r^2+1)r!}{(n+1)!}$$
   for the first few terms, we get 1, 2, 3, 4, 5. This evidence strongly suggests the conjecture that
   $$\sum_{r=1}^{n}(r^2+1)r! = n(n+1)!$$
   and we prove this by induction. For $n = 1$, the conjecture is obviously true, so we assume that it is true for $n = k-1$ (where $k > 1$ is an integer) — that is, that
   $$\sum_{r=1}^{k-1}(r^2+1)r! = (k-1)k!$$

Then we get

$$\sum_{r=1}^{k}(r^2+1)r! = (k-1)k! + (k^2+1)k! = (k^2+k)k! = k(k+1)!.$$

So the conjecture is true for $n = k$, and hence by induction for all positive integers n, and we are done.

Alternative Solution:

This problem can be approached from a different angle by seeking a telescoping form for the summand.

We would like to find a function f such that $f(r+1) - f(r) = (r^2+1)r!$. Clearly if we make $f(r) = r!g(r)$, then $r!$ will be a factor of $f(r+1) - f(r)$. In particular, we get $f(r+1) - f(r) = r!((r+1)g(r+1) - g(r))$. So we must find a function g with $(r+1)g(r+1) - g(r) = r^2 + 1$. If we take g to be a linear function, then $(r+1)g(r+1)$ will be of degree two, as required. Solving for the coefficients, we find that $g(r) = r - 1$ satisfies the identity.

So we have $(r^2+1)r! = (r(r+1)!) - ((r-1)r!)$, and hence

$$\sum_{r=1}^{n}(r^2+1)r! = n(n+1)! - (1-1)1! = n(n+1)!.$$

3. Find the lengths of the sides of any triangle in which each of the numbers $\tan A$, $\tan B$ and $\tan C$ is an integer.

Solution:

$C = \pi - A - B$, so

$$\tan C = \tan(\pi - A - B) = -\tan(A+B) = \frac{\tan A + \tan B}{\tan A \tan B - 1}.$$

Multiplying through, we get $\tan A \tan B \tan C = \tan A + \tan B + \tan C$.

We may assume that $\tan A \leq \tan B \leq \tan C$. Then we get

$$\tan A \tan B \tan C = \tan A + \tan B + \tan C \leq 3 \tan C,$$

so that $3 \geq \tan A \tan B$. This means that we must have $\tan A = 1$, and $\tan B \in \{1, 2, 3\}$. But now, we can determine $\tan C$ from the values of $\tan A$ and $\tan B$ using the identity

$$\tan C = \frac{\tan A + \tan B}{\tan A \tan B - 1} = \frac{\tan B + 1}{\tan B - 1}.$$

Substituting $\tan B = 1$ gives an undefined value of $\tan C$, and substituting $\tan B = 3$ gives $\tan B > \tan C$, contradicting our earlier assumption. So the only solution is $\tan B = 2$, giving $\tan C = 3$.

Now that we have determined all of the angles of the triangle, we can determine the side-lengths up to a constant factor. Let a, b and c be the lengths of the sides opposite angles A, B and C respectively. By the sine rule, we have

$$a = 2R\sin A, \quad b = 2R\sin B, \quad c = 2R\sin C.$$

But for all $\theta \in \left(0, \frac{\pi}{2}\right)$,
$\sin(\tan^{-1}\theta) = \theta/(\sqrt{1+\theta^2})$, so we have

$$\sin A = \frac{1}{\sqrt{2}}, \quad \sin B = \frac{2}{\sqrt{5}} \text{ and } \sin C = \frac{3}{\sqrt{10}}.$$

So for any value of R, we get a solution:

$$\{a, b, c\} = \left\{\sqrt{2}R, \frac{4}{\sqrt{5}}R, \frac{3\sqrt{2}}{\sqrt{5}}R\right\}.$$

4. Let S be a set with precisely n elements, where n is a fixed natural number. If a non-empty subset T of S is chosen at random, show that the probability that T has an odd number of elements is greater than the probability that T has an even number of elements.

Solution:

The total number of subsets of S, including the empty set, is known to be 2^n (since each element has two possibilities – either it is in the set or not), so the total number of non-empty sets is simply $2^n - 1$.

The total number of subsets with an odd number of elements is

$$\sum_r \binom{n}{2r+1} = \frac{1}{2}\sum_r \binom{n}{r} - \frac{1}{2}\sum_r (-1)^r \binom{n}{r}.$$

But now, we know that $(1+u)^n = \sum_r u^r \binom{n}{r}$ for any real number u, so we can simplify this expression to

$$\frac{1}{2}\sum_r \binom{n}{r} - \frac{1}{2}\sum_r (-1)^r \binom{n}{r} = \frac{(1+1)^n - (1-1)^n}{2} = 2^{n-1}.$$

Since every non-empty subset must have either an odd or an even number of elements (and cannot have both), the number of subsets with an even number of elements must be simply $2^n - 1 - 2^{n-1} = 2^{n-1} - 1$.

So we see that there are more subsets with an odd number of elements than with an even number, and so if a subset is selected at random it is more likely to have an odd number of elements than an even number. In particular,

$$P(\text{odd}) = \frac{2^{n-1}}{2^n - 1} > \frac{2^{n-1} - 1}{2^n - 1} = P(\text{even}).$$

5. If A and B are the midpoints of adjacent faces of a cubical box and X is a vertex common to both faces, find, with proof, the size of the angle AXB.

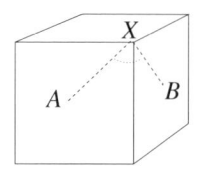

Solution:

Let the cube have side-length $2s$. Then $|XA| = |XB| = \sqrt{s^2 + s^2} = \sqrt{2}s$ and $|AB| = \sqrt{s^2 + s^2} = \sqrt{2}s$. So $\triangle XAB$ is an equilateral triangle, and hence the required angle must be $\pi/3$.

Alternative Solution:

Let the cube have side-length $2s$ once again. We define a set of coordinate axes with the origin at X, and the axes extending along the three sides of the cube, with the z-axis perpendicular to the face containing A, and the y-axis perpendicular to the face containing B.

In this system, we have $\vec{XA} = (s, s, 0)$ and $\vec{XB} = (s, 0, s)$. This allows us to determine the angle between the two vectors, by determining their dot product:

$$\theta = \cos^{-1}\left(\frac{\vec{XA} \cdot \vec{XB}}{|XA||XB|}\right) = \cos^{-1}\left(\frac{s^2}{(\sqrt{2}s)(\sqrt{2}s)}\right) = \cos^{-1}\frac{1}{2} = \frac{\pi}{3}.$$

6. Find, with proof, all solutions of the equation $x! + y! = z!$, where x, y and z are all positive integers.

Solution:

$x!$ is an increasing function from the positive integers to the positive integers, so we must have $z! > z! - x! = y!$, and hence $z > y$. Similarly, $z > x$. We may assume that $y \geq x$.

If $y > x$, then $y, z \geq x+1$, and hence $(x+1)! | (z! - y!) = x!$, which is a contradiction. To avoid contradiction, we must have $x = y$, and hence $2x! = z!$.

But now, $z > x$, so $z \geq 2$ and $(z-1)! \geq x!$. Combining these two inequalities, we get $z(z-1)! \geq 2x!$, with equality if and only if $z = 2$ and $z - 1 = x = 1$.

So the only solution to the equation is

$$(x, y, z) = (1, 1, 2).$$

Alternative Solution:

As before, we must have $z > x, y$. If we divide through by $z!$, we get

$$1 = \frac{x!}{z!} + \frac{y!}{z!} = \frac{1}{\prod_{r=1}^{z-x}(x+r)} + \frac{1}{\prod_{r=1}^{z-y}(y+r)} = \frac{1}{a} + \frac{1}{b}$$

where a and b are positive integers. Now, since $1/a = 1 - 1/b < 1$, we must have $a > 1$, and hence $a \geq 2$ and similarly $b \geq 2$. Finally, this gives $1/a = 1 - 1/b \geq 1/2$, and hence $a \leq 2$, and similarly $b \leq 2$. So we must have $a = b = 2$, and from here it is easy to arrive back at the same unique solution.

7. Find all angles A and B such that $\sin(A+B) = \sin A + \sin B$.

Solution:

If we expand the left hand side using the sine addition rule and gather terms to one side, we get $\sin A(1 - \cos B) + \sin B(1 - \cos A) = 0$.

Now, we can use the double angle formula for cosine to write

$$1 - \cos A = \left(\cos^2 \frac{A}{2} + \sin^2 \frac{A}{2}\right) - \left(\cos^2 \frac{A}{2} - \sin^2 \frac{A}{2}\right) = 2\sin^2 \frac{A}{2}.$$

We can also rewrite $\sin A$ using the double angle formula for sine, to get $\sin A = 2\sin\frac{A}{2}\cos\frac{A}{2}$. Substituting these, we get

$$0 = \left(2\sin\frac{A}{2}\cos\frac{A}{2}\right)\left(2\sin^2\frac{B}{2}\right) + \left(2\sin\frac{B}{2}\cos\frac{B}{2}\right)\left(2\sin^2\frac{A}{2}\right)$$

$$= 4\sin\frac{A}{2}\sin\frac{B}{2}\left(\sin\frac{A}{2}\cos\frac{B}{2} + \cos\frac{A}{2}\sin\frac{B}{2}\right)$$

$$= 4\sin\frac{A}{2}\sin\frac{B}{2}\sin\frac{A+B}{2}.$$

Now, $\sin x = 0$ if and only if $x = n\pi$ for some integer n, so we must have $\sin(A+B) = \sin A + \sin B$ if and only if $A = 2n\pi$, $B = 2n\pi$ or $A + B = 2n\pi$ for some integer n.

8. 131, 181, 15451 are all palindromic primes, i.e. primes equal to the number obtained when the digits in the number are reversed. Find all palindromic primes with an even number of digits.

Solution:

After factoring a few palindromic numbers with an even number of digits, we are quickly led to the conjecture that 11 is always a divisor of such numbers. If this is the case, then we can prove it using modular arithmetic.

Suppose that the number m has $2n$ digits, and is palindromic. Then the digit appearing in position i will also appear in position $2n+1-i$.

If the digit a_i appears in position i for each $1 \leq i \leq n$, then we must have

$$m = \sum_{i=1}^{n} a_1\left(10^i + 10^{2n+1-i}\right).$$

But now, $10^i + 10^{2n+1-i} \equiv (-1)^i + (-1)^{2n+1-i} \equiv 0 \mod 11$, since i and $2n+1-i$ have opposite parities. So we must have $m \equiv 0 \mod 11$, as conjectured.

But if all palindromic primes with an even number of digits must be multiples of 11, then 11 can be the only such prime.

9. The tangent to the curve $y(1+x^2) = 2$ at the point $P = (2, 2/5)$ meets the curve again at Q. Find the coordinates of Q.

Solution:

The slope of the tangent is

$$\frac{dy}{dx}\bigg|_{x=2} = \frac{d}{dx}\left(\frac{2}{1+x^2}\right)\bigg|_{x=2} = \frac{-4x}{(1+x^2)^2}\bigg|_{x=2} = -\frac{8}{25},$$

so its equation is $y - \frac{2}{5} = -\frac{8}{25}(x-2)$, or $y = -\frac{8x+26}{25}$.

This line intersects the curve $y = \frac{2}{1+x^2}$ when $\frac{2}{1+x^2} = -\frac{8x+26}{25}$.
Rearranging, we get $0 = (4x+13)(1+x^2) - 25 = 4x^3 + 13x^2 + 4x - 12$.
Now, we know that $x = 2$ is a double root of this equation, since the line and the curve are tangent at that point. Dividing out this factor, we are left with the root corresponding to the intersection with Q.

$$\frac{4x^3 + 13x^2 + 4x - 12}{(x-2)^2} = (4x+3),$$

so we must have $Q_x = -3/4$.

Putting this value into the equations for either one of the curves gives us the y coordinate of Q. The final answer is $Q = \left(-\frac{3}{4}, \frac{32}{25}\right)$.

10. What, in your opinion, is the next term in the sequence 3, 11, 37, 101, 41, 7, 239,..., and why?

Solution:

We observe firstly that each term in the sequence is a prime. Moreover, if we take the reciprocals of the terms in the series, we see that

$$\frac{1}{3} = 0.\overline{3}, \quad \frac{1}{11} = 0.\overline{09}, \quad \frac{1}{37} = 0.\overline{027}, \quad \frac{1}{101} = 0.\overline{0099}, \quad \frac{1}{41} = 0.\overline{02439},$$

$$\frac{1}{7} = 0.\overline{142857}, \quad \frac{1}{239} = 0.\overline{0041841},$$

so the nth term has a repeating decimal expansion of period n. Moreover, a careful check reveals that in each case, the nth term is the smallest prime number with a repeating decimal expansion of period n. So, for the 8th term, we seek the smallest prime number with a repeating decimal expansion of period 8.

A direct check of the smaller prime numbers reveals that the desired answer is 73, with

$$\frac{1}{73} = 0.\overline{01369863}.$$

Eleventh Superbrain, 1994 — Solutions

1. Given a square in the plane, show how to cut it into four pieces so that these pieces can be reassembled to form two squares of different sizes.

   Solution:

   The simplest way to dissect a square is along a uniformly spaced grid, parallel to its sides. In order to do this dissection in such a manner, we would need each of the three squares to have a side-length of an integer number of units. But since the sum of the areas of the smaller two squares is the area of the larger square, the side-lengths must be a Pythagorean triple.

   Since $\{3, 4, 5\}$ is the smallest integer Pythagorean triple, we might try to find a dissection of a 5×5 square into a 4×4 square and a 3×3 square. It turns out that this is not hard to do:

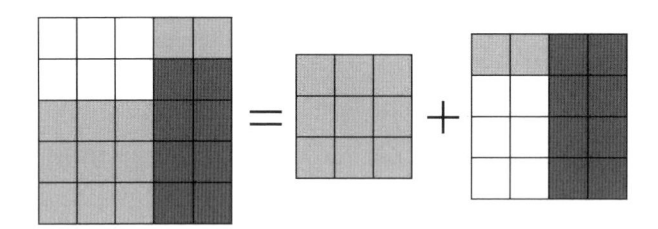

   Remark:

   Without the last condition that the two squares have different sizes, the problem has an even simpler solution:

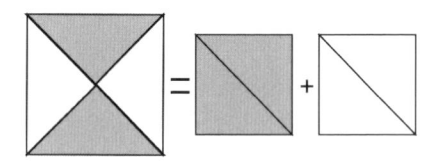

2. If n is a positive integer, show that $n(n+1)(n+2)$ is not the cube of an integer.

   Solution:

   Clearly $n^3 < n^3 + 3n^2 + 2n = n(n+1)(n+2) < n^3 + 3n^2 + 3n + 1 = (n+1)^3$. But now, since $f(n) = \sqrt[3]{n}$ is an increasing function, we may take cube roots across the inequality, to get

   $$n < \sqrt[3]{n(n+1)(n+2)} < n+1.$$

But n and $n+1$ are consecutive integers, and hence there are no integers in the interval $(n, n+1)$. In particular therefore, $\sqrt[3]{n(n+1)(n+2)}$ is not an integer.

3. Evaluate
$$\sum_{n=1}^{\infty} \frac{n-1}{2^{n+1}}.$$

Solution 1:

Adding up the first few partial sums, we get $0, \frac{1}{8}, \frac{1}{4}, \frac{11}{32}, \frac{13}{32}, \frac{57}{128}, \ldots$. Aside from cancellation in a few of the terms, the denominator in the sum of the first n terms is 2^{n+1}. If we write each of the partial sums with 2^{n+1} as the denominator, then we can see the pattern for the numerators:

$$0, 1, 4, 11, 26, 57, 120.$$

Motivated by the presence of the power of two in the question, we observe that the nth term in this sequence is close to 2^n. In fact, a closer inspection reveals that the nth term is $n+1$ less than 2^n for each of the terms calculated.

This leads us to the conjecture that $\sum_{n=1}^{k} \frac{n-1}{2^{n+1}} = \frac{2^k - k - 1}{2^{k+1}}$. We can prove this conjecture by induction on k.

For $k = 1$, the conjecture is certainly true, so we assume that it is true for $k = m$ — i.e. that $\sum_{n=1}^{m} \frac{n-1}{2^{n+1}} = \frac{2^m - m - 1}{2^{m+1}}$. Then we get

$$\sum_{n=1}^{m+1} \frac{n-1}{2^{n+1}} = \frac{2^m - m - 1}{2^{m+1}} + \frac{m}{2^{m+2}} = \frac{2^{m+1} - (m+1) - 1}{2^{m+2}},$$

and so the conjecture is true for $k = m+1$, and hence by induction for all positive integers k.

Finally, $\sum_{n=1}^{\infty} \frac{n-1}{2^{n+1}} = \lim_{k \to \infty} \frac{2^k - k - 1}{2^{k+1}} = \frac{1}{2} - \lim_{k \to \infty} \frac{k+1}{2^{k+1}} = \frac{1}{2}.$

Solution 2:

Let $T_n = (n-1)/(2^{n+1})$. Then as n tends to infinity, T_{n+1}/T_n tends to $1/2$. So, by the ratio test, the series is absolutely convergent. Using this fact, we can obtain a much more direct solution to this problem by avoiding the partial sums, and manipulating the entire series:

Let $S = \sum_{n=1}^{\infty} \frac{n-1}{2^{n+1}}$. Then $2S = \sum_{n=1}^{\infty} \frac{n-1}{2^n} = \frac{0}{2^1} + \sum_{n=1}^{\infty} \frac{n}{2^{n+1}}$. So we have

$$S = 2S - S = \sum_{n=1}^{\infty} \frac{n}{2^{n+1}} - \frac{n-1}{2^{n+1}} = \sum_{n=1}^{\infty} \frac{1}{2^{n+1}} = \frac{1}{2}.$$

Solution 3:

Another direct approach to this problem can be made by using power series. We know that $\sum_{k=0}^{\infty} x^k = \frac{1}{1-x}$ for $x \in (-1, 1)$. Differentiating both sides with respect to x, we get $\sum_{k=0}^{\infty} kx^{k-1} = \frac{1}{(1-x)^2}$. So, setting $x = 1/2$, we have

$$\sum_{n=1}^{\infty} \frac{n-1}{2^{n+1}} = \frac{1}{8}\sum_{n=0}^{\infty} n\left(\frac{1}{2}\right)^{n-1} = \frac{1}{8}\left(\frac{1}{(1-\frac{1}{2})^2}\right) = \frac{1}{2}.$$

4. The sum of a number of positive integers is 100. Find the greatest value their product can have.

Solution:

If we can write a positive integer n as a sum of smaller numbers whose product is greater than n, then we will know that n cannot be one of the integers in question (since we could obtain a larger product by replacing n with these numbers). A few experiments suggest that this can be done for any number greater than 4.

This conjecture is easily confirmed. In fact, given $n > 4$, $n = 2+(n-2)$, but $2(n-2) = n + (n-4) > n$. So we know that none of the integers in question are greater than 4. Moreover, we may replace all 4s with pairs of 2s, so that all of the integers are less than 4. Also, we know that 1 will not be among the numbers, since adding that 1 to any other number would increase the product. So all of the integers may be assumed to be either 2 or 3.

Now, we observe that $2^3 < 3^2$, so that if we have three or more 2s, we may replace three 2s with two 3s, and increase the product. So the number of 2s is at most two.

Finally, if there are a 2s and b 3s, then $2a + 3b = 100$, so that we have $2a \equiv 1 \bmod 3$. Now since $2a \in \{0, 2, 4\}$, we must have $2a = 4$, or $a = 2$, which gives $b = 32$.

So the maximum value of the product is $4 \cdot 3^{32}$.

5. Let y be a function of x and x a function of y, and let $y_0 = y(x_0)$ for some x_0. If $\frac{d^2y}{dx^2}\Big|_{x_0} = 0$, find $\frac{d^2x}{dy^2}\Big|_{y_0}$.

   Solution:

   We know that $\frac{dx}{dy} = \frac{dy}{dx}^{-1}$, so $\frac{d^2x}{dy^2} = \frac{d}{dy}\left(\frac{dy}{dx}^{-1}\right)$. But now, by the chain rule, $\frac{df}{dy} = \frac{dx}{dy}\frac{df}{dx}$ for any function f of x (and hence also of y). So we get

   $$\frac{d^2x}{dy^2} = \frac{dx}{dy}\frac{d}{dx}\left(\frac{dy}{dx}^{-1}\right) = \frac{dx}{dy}\left(-\frac{dy}{dx}^{-2}\right)\left(\frac{d^2y}{dx^2}\right) = -\left(\frac{dx}{dy}\right)^3\frac{d^2y}{dx^2},$$

   since $\frac{d}{dx}\left(\frac{1}{f}\right) = -\frac{1}{f^2}\frac{df}{dx}$ for any function f of x.

   But now, $\frac{d^2y}{dx^2}\Big|_{x_0} = 0$, so we simply get $\frac{d^2x}{dy^2}\Big|_{y_0} = -\left(\frac{dx}{dy}\right)^3\Big|_{y_0}(0) = 0$.

6. If you were given a sequence which began 1, 10, 19, 82, 148, 187, 208, 325, 346, 565, what would you choose as the next term, and why?

   Solution:

   This problem was designed to be almost unsolvable on the night of the test, and the authors are not aware of any systematic method by which the problem can be solved. As such, we merely present the answer (or more precisely, the intended answer — for there may always be others).

   The sequence is the sequence of numbers n such that $10n + 1$, $10n + 3$, $10n + 7$ and $10n + 9$ are all prime. So, 1 is in the sequence since 11, 13, 17 and 19 are all prime; 10 is in the sequence because 101, 103, 107 and 109 are all prime, and so on. The next term is 943.

   It is believed that this sequence is infinite, but to prove this claim would be very difficult — in particular, it is a much stronger generalization of the well-known twin prime conjecture.

7. Solve the equations $x^2 - yz = 1$, $y^2 - zx = 2$, $z^2 - xy = 3$ and verify your solutions.

   Solution:

   Subtracting the first equation from the second, we get $y^2 - x^2 + yz - zx = (y-x)(x+y+z) = 1$. Subtracting the second equation from the third, we get $z^2 - y^2 + zx - xy = (z-y)(x+y+z) = 1$. So we have

   $$y - x = \frac{1}{x+y+z} = z - y.$$

   113

Let this common value be δ. Then we have $\{x, y, z\} = \{y-\delta, y, y+\delta\}$. This gives us $y^2 - zx = y^2 - (y-\delta)(y+\delta) = y^2 - (y^2 - \delta^2) = \delta^2 = 2$, and hence $\delta = \pm\sqrt{2}$.

Finally, returning to $1/(x+y+z) = \delta$, we get

$$y = \frac{x+y+z}{3} = \frac{1}{3\delta} = \pm\frac{1}{3\sqrt{2}}.$$

So there are two solution sets to the equation:

$$\{x_1, y_1, z_1\} = \left\{\frac{1}{3\sqrt{2}} - \sqrt{2}, \frac{1}{3\sqrt{2}}, \frac{1}{3\sqrt{2}} + \sqrt{2}\right\} = \left\{-\frac{5}{3\sqrt{2}}, \frac{1}{3\sqrt{2}}, \frac{7}{3\sqrt{2}}\right\},$$

and $\{x_2, y_2, z_2\} = \{-x_1, -y_1, -z_1\}$.

Alternative Solution:

Multiplying the equations containing x^2, y^2 and z^2 by x, y and z respectively and adding the three, we get

$$x^3 + y^3 + z^3 - 3xyz = x + 2y + 3z.$$

Adding the three directly, we get

$$x^2 + y^2 + z^2 - xy - yz - zx = 6.$$

But now, by a well-known factorization, the left-hand side below divides the left-hand side above:

$$x^3 + y^3 + z^3 - 3xyz = (x+y+z)(x^2 + y^2 + z^2 - xy - yz - zx).$$

So, we get $6(x+y+z) = x + 2y + 3z$, or $5x + 4y + 3z = 0$.

We can use this to substitute in to the original three equations for z, and get three simultaneous equations in x^2, y^2 and xy. These can be solved for x^2 and y^2, and hence for x and y (noting that the signs of the two are linked by the values of xy), with z following from the linear equation.

8. Evaluate

$$\int \frac{dx}{e^x + 1}.$$

Solution:

A good first approach to any integration problem involving a ratio of functions is to make a change of variable to simplify the base. In this case, we might set $u = e^x + 1$. This gives $du = e^x dx$, or $dx = \frac{du}{u-1}$, so the integral becomes

$$\int \frac{dx}{e^x + 1} = \int \frac{du}{u(u-1)}.$$

Using partial fractions, we may write

$$\frac{1}{u(u-1)} = \frac{1}{u-1} - \frac{1}{u},$$

which gives

$$\int \frac{du}{u(u-1)} = \int \left(\frac{1}{u-1} - \frac{1}{u} \right) du = \ln(u-1) + \ln u + C.$$

Replacing u with $e^x + 1$, we get the final answer:

$$\int \frac{dx}{e^x + 1} = \ln(e^x) - \ln(e^x + 1) + C = x - \ln(e^x + 1) + C,$$

where C is an arbitrary constant.

Alternative Solution:

If we divide above and below by e^x, we get an even quicker solution:

$$\int \frac{e^{-x}}{1 + e^{-x}} dx = -\int \frac{d(1 + e^{-x})}{1 + e^{-x}} = -\ln\left(1 + e^{-x}\right) + C,$$

where C is an arbitrary constant.

9. For any set S with exactly n elements, let ${}^n J_r$ be the total number of subsets of S which contain *at least r* elements. Prove that

$$ {}^n J_r + {}^n J_{r-1} = {}^{n+1} J_r \text{ for } 0 \leq r \leq n. $$

Solution:

The definition and the identity are strongly reminiscent of the definition of ${}^n C_r = \binom{n}{r}$ and of the identity $\binom{n}{r} + \binom{n}{r-1} = \binom{n+1}{r}$, and the easiest way to solve this problem is to mimic the classical combinatorial proof of this identity.

Suppose that we want to choose a subset containing at least r elements (with $0 \leq r \leq n$) from a set containing $n + 1$ elements, and suppose that we mark one particular element in the set. When we are choosing a subset with at least r elements, we either include this particular element, or we do not.

If we include this element, then there are $^nJ_{r-1}$ ways to choose at least $r-1$ more elements from the remainder of the set. If we do not include the element, then there are ^nJ_r ways to choose at least r elements from the remainder of the set. So the total number of ways to choose a subset containing at least r elements from a set containing $n+1$ elements must be $^nJ_{r-1} + {^nJ_r}$.

But now, by definition the total number of ways to choose a subset containing at least r elements from a set containing $n+1$ elements is simply $^{n+1}J_r$. So the two expressions for this number must be equal, and we get $^nJ_r + {^nJ_{r-1}} = {^{n+1}J_r}$, exactly as required.

10. In an acute-angled triangle ABC, prove that

$$\tan A \tan B \tan C \geq 3\sqrt{3}.$$

Solution:

$C = \pi - A - B$, so

$$\tan C = \tan(\pi - A - B) = -\tan(A+B) = \frac{\tan A + \tan B}{\tan A \tan B - 1}.$$

Multiplying through, we get

$$\tan A \tan B \tan C = \tan A + \tan B + \tan C.$$

Now, tan is a convex function on $\left(0, \frac{\pi}{2}\right)$, since its second derivative $\frac{d^2}{dx^2} \tan x = 2\sec^2 x \tan x$ is positive for $x \in \left(0, \frac{\pi}{2}\right)$. So, since each of A, B and C lies between zero and $\frac{\pi}{2}$, we get

$$\begin{aligned}\tan A \tan B \tan C &= \tan A + \tan B + \tan C \\ &\geq 3 \tan\left(\frac{A+B+C}{3}\right) \\ &= 3 \tan \frac{\pi}{3} = 3\sqrt{3}.\end{aligned}$$

Twelfth Superbrain, 1995 — Solutions

1. In a hundred metres race, runner A, B and C all run at uniform speeds. If A beats B by 10 metres, and B beats C by 10 metres, by how many metres does A beat C?

   Solution:

   This is another trick question, designed to catch people who give an immediate intuitive answer. It seems natural to conclude that A beats C by 20 metres, but this is not actually the case.

   Suppose that A runs at a speed v_a, B at v_b and c at v_c. Then $v_b = 0.9v_a$ and $b_c = 0.9v_b = 0.81v_a$. So when A has run 100 metres (i.e. finished the race), C has run 81 metres. So A beats C by 19 metres.

   The key thing to remember is that B is not constantly ahead of C by 10 metres — the lead accrues throughout the race. So when A finishes the race and B is only 90% of the way to the finish, B has only accrued 90% of their lead over C. So when A is ten metres ahead of B, B is only nine metres ahead of C.

2. If p, q, r and s are positive real numbers, prove that

   $$\left(\frac{p^2+1}{q}\right)\left(\frac{q^2+1}{r}\right)\left(\frac{r^2+1}{s}\right)\left(\frac{s^2+1}{p}\right) \geq 16.$$

   Solution:

   The way this inequality is written suggests that we have to find some inter-relationship between the variables. However, rewriting it in an equivalent form makes the method of solution a good deal clearer:

   $$\left(\frac{p^2+1}{p}\right)\left(\frac{q^2+1}{q}\right)\left(\frac{r^2+1}{r}\right)\left(\frac{s^2+1}{s}\right) \geq 16.$$

   Now, there are four symmetrical terms in the product, and $16 = 2^4$, so it would suffice to show that each term is not less than two.

   There are many ways to prove that

   $$\frac{x^2+1}{x} = x + \frac{1}{x} \geq 2,$$

   for positive x. For instance, we may use differential calculus to find the minimum of $f(x) = x + \frac{1}{x}$, we may factor

   $$\frac{x^2+1}{x} - 2 = \frac{(x-1)^2}{x},$$

or we may apply the arithmetic-geometric mean inequality, to get

$$x + \frac{1}{x} \geq 2\sqrt{x \cdot \frac{1}{x}} = 2.$$

In any case, applying this result to the original inequality gives

$$\left(\frac{p^2+1}{p}\right)\left(\frac{q^2+1}{q}\right)\left(\frac{r^2+1}{r}\right)\left(\frac{s^2+1}{s}\right) \geq 2.2.2.2 = 16.$$

3. Evaluate

$$\int \frac{\sin x}{\sin x + \cos x} dx.$$

Solution:

If we let $u = \sin x + \cos x$, then we get $du = (\cos x - \sin x)dx$. So we can integrate $\frac{\cos x - \sin x}{\cos x + \sin x}$ with respect to x.

But we can obviously also integrate $1 = \frac{\cos x + \sin x}{\cos x + \sin x}$ with respect to x, so we can evaluate the given integral by combining the two:

$$\int \frac{\sin x}{\sin x + \cos x} dx = \frac{1}{2}\int \left(1 - \frac{\cos x - \sin x}{\sin x + \cos x}\right) dx$$

$$= \frac{x}{2} - \frac{1}{2}\int \frac{du}{u} = \frac{x}{2} - \frac{\ln(\cos x + \sin x)}{2} + C,$$

where C is an arbitrary constant.

4. Let $f(x) = ax^3 + bx^2 + cx + d$ be a function $\mathbb{R} \to \mathbb{R}$, where a, b, c and d are fixed real numbers with $a > 0$. Find necessary and sufficient conditions for the inverse function $f^{-1} : \mathbb{R} \to \mathbb{R}$ to exist.

Solution:

The range of any cubic function is $\mathbb{R}$, so the inverse function will exist if and only if f is strictly increasing or decreasing (and hence one-to-one). Since $a > 0$, f can be strictly monotone only if it is strictly increasing (since $\lim_{x \to -\infty} f(x) = -\infty$ and $\lim_{x \to \infty} f(x) = \infty$).

For f to be strictly increasing, its derivative must be everywhere non-negative. Note that it may be zero at one point — for example, $g(x) = x^3$ is strictly increasing, but $g'(0) = 0$.

$f'(x) = 3ax^2 + 2bx + c$, and since $a > 0$, this is a minimum quadratic. So it is everywhere non-negative if and only if its discriminant is non-negative — i.e. if and only if $(2b)^2 - 4(3a)(c) = 4(b^2 - 3ac) \geq 0$. So a necessary and sufficient condition for the inverse function to exist is

$$b^2 \geq 3ac.$$

5. Let $2, 3, 5, 7, 11, \ldots, p_n, \ldots$ be the sequence of prime numbers. Show that
$$[2.3.5.7.11\ldots p_n] + 1$$
is never a square number.

Solution:

Let $q_n = 1 + \prod_{r=1}^{n} p_r$. Obviously $q_1 = 3$ is not a square, so we assume that $n > 1$. But now, $p_1 = 2$ is the only even prime, so $\prod_{r=2}^{n} p_r$ must be an odd number. So we may write $\prod_{r=2}^{n} p_r = 2k + 1$ for some integer k. But then, $q_n = \prod_{r=1}^{n} p_r + 1 = 4k + 3$.

But every square of an integer is congruent to either zero or one modulo four, so $q_n \equiv 3 \mod 4$ means that q_n cannot be a square for any positive integer n.

6. Evaluate
$$\sum_{n=2}^{\infty} \log_2\left(1 - \frac{1}{n^2}\right).$$

Solution:

$\log_2\left(1 - \frac{1}{n^2}\right) = \log_2\left(\frac{(n+1)(n-1)}{(n)(n)}\right) = \log_2\frac{n+1}{n} - \log_2\frac{n}{n-1}$, giving a telescoping form of the summand. So we can write

$$\sum_{n=2}^{k} \log_2\left(1 - \frac{1}{n^2}\right) = \sum_{n=2}^{k}\left(\log_2\frac{n+1}{n} - \log_2\frac{n}{n-1}\right) = \log_2\frac{k+1}{k} - \log_2 2.$$

So $\sum_{n=2}^{\infty} \log_2\left(1 - \frac{1}{n^2}\right) = \lim_{k\to\infty}\left(\log_2\left(1 + \frac{1}{k}\right) - 1\right).$

But the log function is continuous at 1, so

$$\lim_{k\to\infty} \log_2\left(1 + \frac{1}{k}\right) = \log_2\left(\lim_{k\to\infty} 1 + \frac{1}{k}\right) = \log_2(1) = 0.$$

Substituting this into the above expression for the sum, we get

$$\sum_{n=2}^{\infty} \log_2\left(1 - \frac{1}{n^2}\right) = -1.$$

7. A lampshade is in the shape of a frustrum of a right-circular cone with r the radius of the top, R the radius of the base and l the slant height. Find a formula for the area of the curved surface of the lampshade.

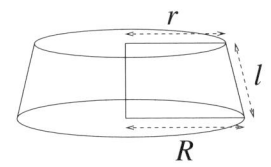

Solution:

We extend the curved surface area of the frustrum to form a complete cone. Let θ be the apex angle of the completed cone. Then we must have $\sin\theta = (R-r)/l$.

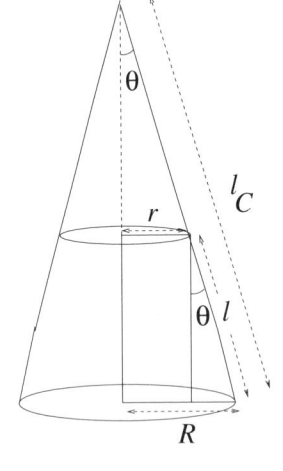

So the slant height of the complete cone is

$$l_C = \frac{R}{\sin\theta} = \frac{Rl}{R-r}.$$

The slant height of the piece added to the frustrum to complete the cone is $l_C - l = rl/(R-r)$. The curved surface area of a circular cone is half the perimeter of the base times the slant height (note what happens in the limit as a cone flattens to a disc). Now $\pi R l_C = (\pi R^2 l)/(R-r)$, while the curved area of the section added is $\pi r(l_C - l) = \pi r^2 l/(R-r)$.

The curved surface area of the frustrum is simply the difference of these two, or

$$\frac{\pi R^2 l}{R-r} - \frac{\pi r^2 l}{R-r} = \pi l \frac{R^2 - r^2}{R-r} = \pi l (R+r).$$

8. Without calculating either number, decide which of the numbers π^e or e^π is the greater. [You may assume that $\pi > e$, where π is the ratio of the length of the circumference of a circle to its diameter, and e is the base of natural logarithms.]

Solution:

Since we are asked to work with very little information about either number, we consider what properties either one has which might help in solving this problem. Since the properties of π relate (generally speaking) to circles and their properties, they are not of much use in this question. However, one of the defining features of e is that $\frac{d}{dx}e^x = e^x$ — so if we can construct a function whose critical points are of significance to determining which of the two numbers is the larger, then we could utilise this property.

For $x > 0$, define $f(x) = e^x/x^e$. Then $f'(x) = \dfrac{e^x}{x^e} - \dfrac{e^{x+1}}{x^{e+1}} = \dfrac{e^x}{x^{e+1}}(x-e)$. Now, since $e^x/(x^{e+1})$ is positive for all $x > 0$, we must have $f'(x) < 0$ for $x < e$, and $f'(x) > 0$ for $x > e$. So, since it is differentiable on $(0, \infty)$, f decreases on $(0, e)$ and increases on (e, ∞). So f must attain its absolute minimum on $(0, \infty)$ at the point $x = e$. So we have $\dfrac{e^\pi}{\pi^e} = f(\pi) > f(e) = \dfrac{e^e}{e^e} = 1$ and hence $e^\pi > \pi^e$.

9. A plane quadrilateral has sides of length a, b, c and d. If
$$a^2 + b^2 + c^2 + d^2 = ab + bc + cd + da,$$
prove that the diagonals of the figure are perpendicular to each other.

Solution:

Although this problem appears in a geometric form, it is in fact an inequality in disguise. If the solver is familiar with the rearrangement inequality, then they will immediately recognise that the left hand side of the equation in the side-lengths is greater than or equal to the right hand side, with equality if and only if $a = b = c = d$.

This fact can be proved by less sophisticated means. We can write the difference between the two sides as a sum of squares:

$$0 \leq \frac{1}{2}\left((a-b)^2 + (b-c)^2 + (c-d)^2 + (d-a)^2\right)$$

$$= a^2 + b^2 + c^2 + d^2 - ab - bc - cd - da.$$

Clearly we have equality if and only if each of the terms being squared is zero, and hence if and only if $a = b = c = d$.

So, since we are given $a^2 + b^2 + c^2 + d^2 = ab + bc + cd + da$, it follows that $a = b = c = d$. So the quadrilateral in question must a rhombus. But it is easy to show that the diagonals of a rhombus are always perpendicular to one another, and so the result is proved.

10. Let ABC be a triangle and r be the radius of the circle which touches all three sides of the triangle internally. Let p_1, p_2 and p_3 be the perpendicular distances of each of the vertices of the triangle to the opposite side (produced if necessary). Show that

$$\frac{1}{r} = \frac{1}{p_1} + \frac{1}{p_2} + \frac{1}{p_3}.$$

Solution:

Let a_i be the side whose corresponding altitude is p_i, and let the area of the triangle be Δ. Then $\frac{1}{2} a_i p_i = \Delta$. Rearranging, we get $\frac{1}{p_i} = \frac{a_i}{2\Delta}$. So

$$\frac{1}{p_1} + \frac{1}{p_2} + \frac{1}{p_3} = \frac{a_1 + a_2 + a_3}{2\Delta}. \tag{1}$$

If we join the incentre to each of the three vertices, we partition the triangle into three smaller triangles, each with one of the sides of the main triangle. Let Δ_i be the area of the triangle which has the side of length a_i as one of its sides.

Since each of the three triangles has an altitude of r (to the apex at the incentre) above the base side, we must have $\Delta_i = \frac{1}{2} a_i r$. But since these three partition the main triangle, we must have $\Delta = \Delta_1 + \Delta_2 + \Delta_3 = \frac{1}{2}(a_1 + a_2 + a_3)r$.

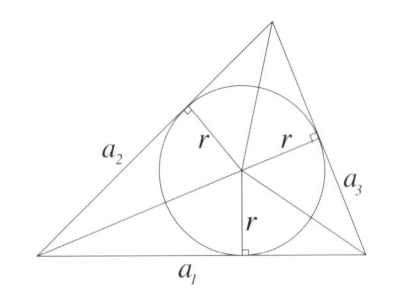

Substituting this into (1), we get

$$\frac{1}{p_1} + \frac{1}{p_2} + \frac{1}{p_3} = \frac{a_1 + a_2 + a_3}{2\Delta} = \frac{a_1 + a_2 + a_3}{(a_1 + a_2 + a_3)r} = \frac{1}{r}.$$

Thirteenth Superbrain, 1996 — Solutions

1. Which is a closer fit, a square peg in a round hole or a round peg in a square hole?

   Solution:

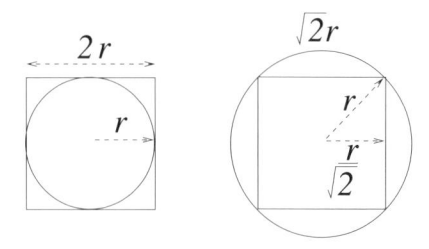

   If we circumscribe a square around a circle of radius r, then it will have sides of length $2r$, and an area of $4r^2$. So the proportion of the square filled by the circle is
   $$\frac{\pi r^2}{4r^2} = \frac{\pi}{4}.$$

   On the other hand, if we inscribe a square in a circle of radius r, then it will have sides of length $\sqrt{2}r$, and an area of $2r^2$. So the proportion of the circle filled by the square is
   $$\frac{2r^2}{\pi r^2} = \frac{2}{\pi}.$$

   But now, $\pi > 3$, so $\pi^2 > 8$, and we get $\pi/4 > 2/\pi$. So the round peg fills a greater proportion of the square hole than the square peg fills of the round hole.

2. Using coins with face value 1p, 2p, 5p, 10p, 20p and 50p, and these only, what is the largest amount of money that you could have and not be able to make up £1 (100p) exactly?

   Solution:

   Suppose that we have some set of coins, with which we cannot make up £1 exactly.

   If there is more than one 1p coin in the set, then we may group pairs of them together, and treat each of the pairs as a 2p coin. So we may assume that there is at most one 1p coin. Similarly, if there are more than five 2p coins, we may group them in sets of five, and treat each of the sets as a 10p coin, so that we may assume that there are at most four 2p coins.

Continuing in this manner, we may assume that there is not more than one 5p coin, not more than one 10p coin, not more than four 20p coins, and not more than one 50p coin.

If we assume that each of the individual limits are met, then we have a total of 1×1p+4×2p+1×5p+1×10p+4×20p+1×50p=£1.54. However, with this set, we can still make £1 exactly.

If there are two or more 2p coins in the set, and one 1p coin, then we may group these together and treat them as a 5p coin. So we may assume that either there are no 1p coins, or there is at most one 2p coin. Similarly, either there is no 10p coin, or there is at most one 20p coin.

With these additional constraints, the maximum value that the set can have is 4×2p+1×5p+4×20p+1×50p=£1.43. So any set of coins from which £1 cannot be made up exactly has a value of at most £1.43. We claim that with this set, it is impossible to make up £1 exactly. We prove this by contradiction.

If some subset of this set has a value of £1 exactly, then the complement of that set has a value of 43p. In order to have an odd total value, the complement must include the 5p coin. So the remainder of the set must have a value of 38p, and hence must contain all of the 2p coins. But now, we must make up 30p from 20p and 50p coins, which is impossible.

3. If $x = \sqrt{2} + \sqrt{3}$, evaluate
$$x^8 - 98x^4 + x + 17.$$

Solution:

This problem is simply an extended calculation. However, multiplying out $(\sqrt{2} + \sqrt{3})^8$ directly is a tedious calculation, and it is quite easy to make a mistake. It is more efficient and more reliable to calculate x^8 by repeated squaring (since $x^8 = (x^4)^2$, $x^4 = (x^2)^2$).

Proceeding in this manner, we get $x^2 = 5 + 2\sqrt{6}$ and $x^4 = 49 + 20\sqrt{6}$.

Now, we can simplify the calculations even more by observing that
$$\begin{aligned}x^8 - 98x^4 &= x^4(x^4 - 98) = (20\sqrt{6} + 49)(20\sqrt{6} - 49) \\ &= (20\sqrt{6})^2 - 49^2 = 2400 - 2401 = -1.\end{aligned}$$

Substituting this value into the expression, we get
$$x^8 - 98x^4 + x + 17 = 16 + x = 16 + \sqrt{2} + \sqrt{3}.$$

4. The plane is to be tiled with regular n-sided polygons, all of the same shape and size. Find all possible values that n can have.

Solution:

Tiling with a regular n-gon for $n = 3, 4$ or 6 is quite simple:

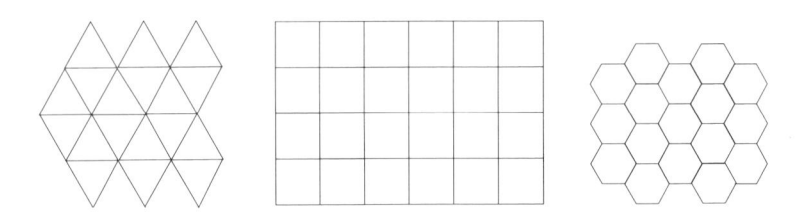

For $n = 5$ or $n > 6$, a little experimentation should serve to convince one that such a tiling is not possible.

Suppose that the plane were tiled with regular n-gons for one of these n. Consider a vertex of one of the figures. For a small neighbourhood around this point to be covered, there must be one or more other figures incident on this point. If they cover the neighbourhood, then the angles which are incident on the point must add up to at least 2π. But if the figures are not to intersect, then the sum must be at most 2π.

So if the tiling satisfies the properties in the question, then the sum of the angles incident on the point is exactly 2π. But now, the only angles in a regular n-gon are π (on a side) and $\pi - \frac{2\pi}{n}$ (at the vertices). So we must have a sum of numbers of this form equal to 2π. Also, since the point in question is the vertex of one figure, at least one of these is not π.

For $n > 6$, every angle θ of the n-gon satisfies $\pi \geq \theta > \frac{2\pi}{3}$. So it is not possible for three figures to be incident on a point, since this would give a sum greater than 2π. But if only two figures are incident on the point, and the incident angle for one of them is $\pi - \frac{2\pi}{n} < \pi$, then the sum of the two angle must be less than 2π. So it is not possible for the given neighbourhood to be covered for $n > 6$.

Now let $n = 5$. Then we need to find a solution to the equation $a\frac{3\pi}{5} + b\pi = 2\pi$, or $3a + 5b = 10$, where $a > 0$ and $b \geq 0$ are integers. So, we must have $5|(10 - 5b) = 3a$, and hence $5|a$. But $10 - 3a = 5b \geq 0$, so we must have $a \in \{1, 2, 3\}$, and hence the congruence cannot be satisfied.

So there are no solutions to the equation, and hence the tiling is not possible with $n = 5$. So the only possible values of n are 3, 4 and 6.

5. Evaluate
$$\sum_{r=1}^{n} \sin rx.$$

Solution:

If $\sin x = 0$ then $x = k\pi$ for some integer k, and hence $\sin rx = \sin(rk)\pi = 0$ for all positive integers r, and the sum is equal to zero. Otherwise, we can use the identity $2\sin A \sin B = \cos(A-B) - \cos(A+B)$ to write

$$\sin rx = \frac{1}{2\sin x}\left(\cos(r-1)x - \cos(r+1)x\right).$$

Let $S = \sum_{r=1}^{n} \sin rx$. Then we have

$$\begin{aligned}
2S \sin x &= \cos 0 - \cancel{\cos 2x} \\
&\quad + \cos x - \cancel{\cos 3x} \\
&\quad + \cancel{\cos 2x} - \cancel{\cos 4x} \\
&\quad + \ldots \\
&\quad + \cancel{\cos(n-2)x} - \cos nx \\
&\quad + \cancel{\cos(n-1)x} - \cos(n+1)x \\
&= 1 + \cos x - \cos nx - \cos(n+1)x,
\end{aligned}$$

and hence $S = \dfrac{1}{2\sin x}\left(1 + \cos x - \cos nx - \cos(n+1)x\right).$

6. If x, y and z are positive integers with $x \leq y \leq z$, find all solutions of the equation
$$xyz = x + y + z.$$

Solution:

Applying the inequalities given to the equation, we get
$$xyz = x + y + z \leq 3z,$$

and hence $xy \leq 3$. Since $x \leq y$, we must have $x = 1$. Substituting this into the original equation gives
$$yz = 1 + y + z, \quad \text{or} \quad (y-1)(z-1) = 2.$$

But $y - 1 \leq z - 1$, so we must have $y - 1 = 1$, $z - 1 = 2$. So the only solution is
$$(x, y, z) = (1, 2, 3).$$

7. Both 30716 and 65419378 are natural numbers with the property that no digit is repeated. How many such natural numbers are there?

   Solution:

   Given any set of k distinct non-zero digits, we can make $k!$ different numbers by rearranging these digits. We can also make another $k.k!$ numbers by substituting a zero after digit i of each of these numbers for $1 \leq i \leq k$. So we can make a total of $(k+1)!$ different natural numbers with no digit repeated for each positive integer k and each set of k distinct non-zero digits.

   Now, since every natural number with no repeated digits has a corresponding set of non-zero digits in the number, we know that every number with the desired property has been counted at least once. But clearly all of the numbers are different to each other, and so each one is counted exactly once.

   Summing over all possible values of k, and over the total number of subsets of size k, we get that the total number of natural numbers with no digit repeated is

   $$\sum_{k=1}^{9} (k+1)! \binom{9}{k} = 8877690.$$

8. Evaluate
   $$\int \frac{dx}{\sin x + \cos x}.$$

   Solution:

   We know that $\sin \frac{\pi}{4} = \cos \frac{\pi}{4} = \frac{1}{\sqrt{2}}$, and so we can write

   $$\sin x + \cos x = \sqrt{2}\left(\sin x \cos \frac{\pi}{4} + \cos x \sin \frac{\pi}{4}\right) = \sqrt{2} \sin\left(x + \frac{\pi}{4}\right).$$

   So the integral becomes $\int \frac{dx}{\sin x + \cos x} = \frac{1}{\sqrt{2}} \int \operatorname{cosec}\left(x + \frac{\pi}{4}\right) dx.$

   But now, we know that $\int \operatorname{cosec} x \, dx = \ln\left|\tan \frac{x}{2}\right|$, so we get

   $$\int \frac{dx}{\sin x + \cos x} = \frac{1}{\sqrt{2}} \ln\left|\tan\left(\frac{x}{2} + \frac{\pi}{8}\right)\right| + C,$$

   where C is an arbitrary constant.

9. Find functions $f(x)$ and $g(x)$ such that

$$\frac{d}{dx}(f(x)g(x)) = \frac{df(x)}{dx}\frac{dg(x)}{dx},$$

where neither $f(x)$ nor $g(x)$ is a constant function.

Solution:

We shall write $f = f(x)$, $f' = f'(x) = \frac{df(x)}{dx}$, and similarly for g and other functions.

By the product rule and from the given equation, we have

$$(fg)' = fg' + gf' = f'g'.$$

If we take $f' \neq 0$ and $g' \neq 0$, then we can divide across by $f'g'$ to get $\frac{f}{f'} + \frac{g}{g'} = 1$. This equation will be satisfied by $f = g = h$ if we can find a function $h(x)$ such that $\frac{h}{h'} = \frac{1}{2}$.

So we have a differential equation in h to solve. Inverting both sides gives us $\frac{h'}{h} = 2$. Integrating both sides of this equation with respect to x, we get

$$\int \frac{1}{h}\frac{dh}{dx}dx = \int \frac{dh}{h} = \ln h = 2x + C,$$

where C is an arbitrary constant.

So $h(x) = e^{2x+C}$ satisfies the equation for any real constant C, and also satisfies the conditions that $h' \neq 0$ and that h is non-constant. So $f(x) = g(x) = e^{2x+C}$ is a solution of the form required.

Remark:

In fact, $f(x) = e^{px+a}$, $g(x) = e^{qx+b}$ is a solution for any $a, b \in \mathbb{R}$, and any $p, q \in \mathbb{R}$ with $\frac{1}{p} + \frac{1}{q} = 1$.

10. Show that the sum of the squares of four consecutive numbers is never a square number.

Solution:

Let the first of the four numbers be n, for some integer n. Then the sum of the four squares is $n^2 + (n+1)^2 + (n+2)^2 + (n+3)^2 = 4n^2 + 12n + 14$.

But now, we know that $m^2 \equiv 0$ or $1 \mod 4$ for all integers m, whereas $4n^2 + 12n + 14 = 4(n^2 + 3n + 3) + 2 \equiv 2 \mod 4$. So $4n^2 + 12n + 14$ cannot be the square of an integer.

Alternative Solution:

The sum of the four squares starting from n is the same as the sum of the four squares starting from $-3-n$, so we need only prove the claim for $n > -2$ (since for $n \leq -2$, $-3-n > -2$). Moreover, for $n = -1$, the sum is 6, which is not a square. So we may assume $n \geq 0$.

Again, we get that this sum is $4n^2 + 12n + 14$. But now,

$$(2n+3)^2 = 4n^2+12n+9 < 4n^2+12n+14 < 4n^2+16n+16 = (2n+4)^2$$

holds for any integer $n \geq 0$. Taking square roots, we get

$$2n+3 < \sqrt{n^2 + (n+1)^2 + (n+2)^2 + (n+3)^2} < 2n+4.$$

But $2n+3$ and $2n+4$ are consecutive integers, and so there are no integers in the interval $(2n+3, 2n+4)$. In particular, the sum of the four squares starting from n is not the square of an integer.

Fourteenth Superbrain, 1997 — Solutions

1. Find all prime numbers p such that $2p - 1$ and $2p + 1$ are also prime numbers.

   Solution:

   It is easy to see that $p = 2$ and $p = 3$ both satisfy the property.

   After testing a number of subsequent values, we are led to the conjecture that these are the only such values. In fact, for each subsequent prime p, we find that one of $2p - 1$ and $2p + 1$ is a multiple of three.

   Since $p = 3$ is the only prime that is a multiple of three, any prime $p > 3$ must be of the form $3k + 1$ or $3k + 2$ for some positive integer k. If $p = 3k + 1$, then $2p + 1 = 6k + 3 = 3(2k + 1)$, and so $2p + 1$ is not prime (since $2k + 1 > 1$). Similarly, if $p = 3k + 2$, then $2p - 1 = 6k + 3 = 3(2k + 1)$, and so $2p - 1$ is not prime.

   So in any case, for $p > 3$, at least one of $2p - 1$ and $2p + 1$ is not prime, and so the only prime numbers p with the given property are $p = 2$ and $p = 3$.

2. A piece of wire 100cm in length is bent into the shape of a sector of a circle. Find the maximum value that the area A of the sector can have.

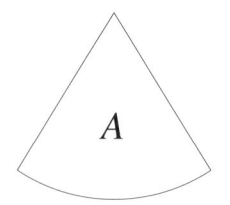

   Solution:

   Let the radius of the sector be r, and the length of the curved portion be l. Then $2r + l = 100$cm.

   The area of the sector is $A = \frac{1}{2}r^2\theta$, where θ is the angle in the sector (measured in radians). But by the definition of the radian, $\theta = \frac{l}{r}$, so we get $A = \frac{1}{2}r^2\left(\frac{l}{r}\right) = \frac{rl}{2}$.

   But now, by the arithmetic geometric mean inequality, $\left(\frac{2r+l}{2}\right)^2 \geq 2rl$, so we get $A = \frac{rl}{2} \leq \frac{(2r+l)^2}{16} = \frac{1}{16}$m².

   The case of equality occurs when the slant height l is twice the radius r, or when the angle in the sector is 2 radians.

   Remark:

   If the reader is not familiar with the AM-GM inequality, the same result can be achieved by substituting in for l to get $A = \frac{rl}{2} = r(50\text{cm} - r)$ and differentiating to find the maximum.

3. A tangent touches the curve $y = x^2(x-1)(x-3)$ at two distinct points. Find the equation of the tangent.

   Solution:

   Let $y = ax + b$ be the equation of the line to which the curve is tangent at two points. Then $f(x) = x^2(x-1)(x-3) - ax - b$ must be tangent to the line $y = 0$ at two distinct points, and hence it must have a double root at each of these points. Let these points be $x = c$ and $x = d$. But since a monic degree four polynomial is determined completely by the locations of its four roots, we must have

   $$f(x) = x^2(x-1)(x-3) - ax - b = (x-c)^2(x-d)^2.$$

   Expanding each of the polynomials, and setting the coefficients of the corresponding powers of x equal to one another, we get four simultaneous equations for a, b, c and d: $4 = 2c + 2d$, $3 = c^2 + 4cd + d^2$, $a = 2c^2d + 2cd^2$ and $b = -c^2d^2$.

   From the first equation, we get $c + d = 2$. Squaring, we get $c^2 + 2cd + d^2 = 4$, and subtracting this from the second equation, we get $cd = -\frac{1}{2}$. Substituting these two values into the third equation, we get $a = 2(c+d)(cd) = 2(2)(-1/2) = -2$. Finally, from the fourth equation, we get $b = -(cd)^2 = -1/4$.

   So we have the values of a and b, and hence we know the equation of the tangent line to the curve: $y = ax + b = -2x - 1/4$.

4. Show that there are exactly three kinds of plane triangle where the number of degrees in each angle is an integer divisor of 180.

   Solution:

   Let the angles of the triangle be A, B, and C, with $A \geq B \geq C > 0$. Let $A = \frac{180°}{n_a}$, $B = \frac{180°}{n_b}$ and $C = \frac{180°}{n_c}$. Then $2 \leq n_a \leq n_b \leq n_c$, and $\frac{1}{n_a} + \frac{1}{n_b} + \frac{1}{n_c} = 1$. Also, $\frac{3}{n_a} \geq \frac{1}{n_a} + \frac{1}{n_b} + \frac{1}{n_c} = 1$, so we must have $n_a = 2$ or 3.

   If $n_a = 2$, then $\frac{1}{n_b} + \frac{1}{n_c} = \frac{1}{2}$, and hence $2 < n_b \leq n_c$. Also, $\frac{2}{n_b} \geq \frac{1}{n_b} + \frac{1}{n_c} = \frac{1}{2}$, so we must have $n_b = 3$ or 4. If $n_b = 3$, then $n_c = 6$, while $n_b = 4$ yields $n_c = 4$, so both cases provide integer solutions.

   Finally, $n_a = 3$ gives $n_b = n_c = 3$, since $n_a = n_b = n_c$ is a necessary condition for equality in $3/n_a \geq 1$.

So there are exactly three solutions in positive integers n_a, n_b and n_c to the equation $1/n_a + 1/n_b + 1/n_c = 1$. Correspondingly, there are exactly three kinds of plane triangle where the number of degrees in each angle is an integer divisor of 180. These are

$$\{60°, 60°, 60°\}, \quad \{90°, 45°, 45°\} \quad \text{and} \quad \{90°, 60°, 30°\}.$$

5. If a and b are positive real numbers with $a > b$, prove that

$$\frac{(a-b)^2}{8a} < \frac{a+b}{2} - \sqrt{ab} < \frac{(a-b)^2}{8b}.$$

Solution:

We can write the centre term as $(a+b)/2 - \sqrt{ab} = (\sqrt{a} - \sqrt{b})^2/2$. We can factor $a - b$ as a difference of two squares. By writing $a - b = (\sqrt{a} - \sqrt{b})(\sqrt{a} + \sqrt{b})$ we can factor out this term from each of the other two parts of the inequality. On the left hand side, we get

$$\frac{(a-b)^2}{8a} = \frac{(\sqrt{a}-\sqrt{b})^2}{2} \cdot \frac{(\sqrt{a}+\sqrt{b})^2}{4a} < \frac{(\sqrt{a}-\sqrt{b})^2}{2}.$$

Now since $\sqrt{a} - \sqrt{b} > 0$ and $a > 0$, we can multiply through by the positive quantity $8a/(\sqrt{a} - \sqrt{b})^2$ to get the equivalent inequality $(\sqrt{a} + \sqrt{b})^2 < 4a$. Since both sides are positive, this is equivalent to $\sqrt{a} + \sqrt{b} < 2\sqrt{a}$, which is true.

Similarly, on the right hand side, we get

$$\frac{(\sqrt{a}-\sqrt{b})^2}{2} < \frac{(a-b)^2}{8b} = \frac{(\sqrt{a}+\sqrt{b})^2}{4b} \cdot \frac{(\sqrt{a}-\sqrt{b})^2}{2}.$$

Multiplying through by the positive quantity $8b/(\sqrt{a} - \sqrt{b})^2$, we get the equivalent inequality $4b < (\sqrt{a} + \sqrt{b})^2$. Since both sides are positive, this is equivalent to $2\sqrt{b} < \sqrt{a} + \sqrt{b}$, which is true.

So both sides of the inequality have been established, and we are done.

6. Evaluate

$$\int_0^\infty \frac{dx}{x^3 + 1}.$$

Solution:

We can rewrite $1/(x^3 + 1)$ using partial fractions as

$$\frac{a}{x+1} + \frac{bx+c}{x^2 - x + 1}.$$

Solving for a, b and c, we get

$$\frac{1}{x^3+1} = \frac{1}{3}\left(\frac{1}{x+1} + \frac{2-x}{x^2-x+1}\right).$$

So

$$\int_0^\infty \frac{dx}{x^3+1} = \frac{1}{3}\lim_{n\to\infty}\left(\int_0^n \frac{dx}{x+1} + \int_0^n \frac{2-x}{x^2-x+1}\right).$$

Now, it is easy to calculate $\int_0^y \frac{1}{x+1}dx = \ln(y+1)$. So we need to evaluate $\int_0^y \frac{2-x}{x^2-x+1}dx$. If we let $u = x^2 - x + 1$, we get $du = 2x - 1$, so we get

$$\int_0^n \frac{\frac{1}{2}-x}{x^2-x+1}dx = -\frac{1}{2}\int_1^{n^2-n+1}\frac{du}{u} = -\frac{1}{2}\ln(n^2-n+1).$$

So now we need only evaluate

$$\int_0^n \frac{\frac{3}{2}dx}{x^2-x+1}.$$

Completing the square in the denominator, we get

$$x^2 - x + 1 = \left(x - \frac{1}{2}\right)^2 + \left(\frac{\sqrt{3}}{2}\right)^2.$$

So, setting $v = x - 1/2$ and $a = \sqrt{3}/2$, we get

$$\int_0^n \frac{\frac{3}{2}dx}{x^2-x+1} = \frac{3}{2}\int_{-\frac{1}{2}}^{n-\frac{1}{2}} \frac{dv}{v^2+a^2} = \frac{3}{2}\left[\frac{\tan^{-1}\frac{v}{a}}{a}\right]_{-\frac{1}{2}}^{n-\frac{1}{2}}$$

$$= \sqrt{3}\left(\tan^{-1}\left(\frac{2n-1}{\sqrt{3}}\right) - \tan^{-1}\left(-\frac{1}{\sqrt{3}}\right)\right)$$

$$= \sqrt{3}\left(\frac{\pi}{6} + \tan^{-1}\left(\frac{2n-1}{\sqrt{3}}\right)\right).$$

With all the component parts evaluated, all we need to do it combine the results, and then take the limit as n goes to infinity. We get

$$\int_0^\infty \frac{dx}{x^3+1} = \frac{1}{3}\lim_{n\to\infty}\left(\ln(n+1) - \frac{1}{2}\ln(n^2-n+1)\right.$$

$$\left. + \sqrt{3}\left(\frac{\pi}{6} + \tan^{-1}\left(\frac{2n-1}{\sqrt{3}}\right)\right)\right).$$

Now, $\lim\limits_{n\to\infty}\left(\ln(n+1)-\dfrac{1}{2}\ln(n^2-n+1)\right) = \lim\limits_{n\to\infty}\ln\left(\dfrac{n+1}{\sqrt{n^2-n+1}}\right)$.

But $\lim\limits_{n\to\infty}\dfrac{n+1}{\sqrt{n^2-n+1}} = \lim\limits_{n\to\infty}\dfrac{1+\frac{1}{n}}{\sqrt{1-\frac{1}{n}+\frac{1}{n^2}}} = 1$, and $\ln(x)$ is continuous at $x=1$, so we get $\lim\limits_{n\to\infty}\ln\left(\dfrac{n+1}{\sqrt{n^2-n+1}}\right) = \ln(1) = 0$.

Finally we have, $\lim\limits_{n\to\infty}\tan^{-1}\left(\dfrac{2n-1}{\sqrt{3}}\right) = \tan^{-1}(\infty) = \dfrac{\pi}{2}$, so that we get $\dfrac{1}{3}\lim\limits_{n\to\infty}\sqrt{3}\left(\dfrac{\pi}{6}+\tan^{-1}\left(\dfrac{2n-1}{\sqrt{3}}\right)\right) = \dfrac{2\pi}{3\sqrt{3}}$.

Combining the two limits, we get
$$\int_0^\infty \dfrac{dx}{x^3+1} = \dfrac{2\pi}{3\sqrt{3}}.$$

Alternative Solution:

Although the systematic approach with partial fractions is effective, we are actually solving a more difficult problem — to find the value of the integral over $[0, n]$ for all n, rather than only its value over $[0, \infty)$. If we use a trick to take into account the limits of integration, then we can obtain a quicker solution.

In particular, we first split the integral as follows:
$$\int_0^\infty \dfrac{dx}{x^3+1} = \int_0^1 \dfrac{dx}{x^3+1} + \int_1^\infty \dfrac{dx}{x^3+1}.$$

Now in the second term, we make the substitution $x = 1/y$, $dx = (-1/y^2)dy$ to get
$$\int_1^\infty \dfrac{dx}{x^3+1} = -\int_1^0 \dfrac{\frac{1}{y^2}}{\frac{1}{y^3}+1}dy = \int_0^1 \dfrac{y}{y^3+1}dy.$$

Substituting this back into the original integral, we get
$$\int_0^\infty \dfrac{dx}{x^3+1} = \int_0^1 \dfrac{y+1}{y^3+1}dy = \int_0^1 \dfrac{dy}{y^2-y+1},$$

which can be integrated in the same way as in the first solution by setting $v = y - 1/2$ and $a = \sqrt{3}/2$ to get
$$\int_0^\infty \dfrac{dx}{x^3+1} = \int_0^1 \dfrac{dy}{y^2-y+1} = \int_{-\frac{1}{2}}^{\frac{1}{2}} \dfrac{dv}{v^2+a^2} = \left[\dfrac{\tan^{-1}\frac{v}{a}}{a}\right]_{-\frac{1}{2}}^{\frac{1}{2}} = \dfrac{2\pi}{3\sqrt{3}}.$$

7. In the given triangle, the areas are as shown. Find the area x of the quadrilateral.

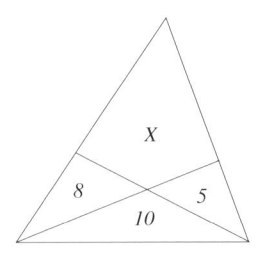

Solution:

Let the vertices of the triangle be A, B and C, let the point of intersection of the two Cevians be O, and let the points where these meet the opposite sides be P and Q (see diagram). Let $\triangle(UVW)$ be the area of the triangle with vertices at U, V and W.

The key to solving this problem is to connect the intersection of the two Cevians of the triangle to the third vertex, thus splitting the quadrilateral X into two sections. We assume that these have areas of x_1 and x_2 respectively.

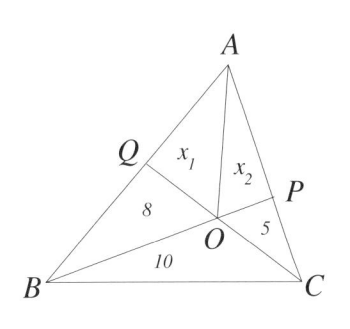

Now, since the area of a triangle standing on a given base is proportional to its height, we must have

$$\frac{\triangle(BCQ)}{\triangle(BOQ)} = \frac{|QC|}{|QO|} = \frac{\triangle(ACQ)}{\triangle(AOQ)}$$

and

$$\frac{\triangle(ABP)}{\triangle(AOP)} = \frac{|PB|}{|PO|} = \frac{\triangle(CBP)}{\triangle(COP)}.$$

Substituting the given values for the areas, we get two simultaneous equations in x_1 and x_2:

$$\frac{15}{5} = \frac{x_1 + x_2 + 8}{x_2} \quad \text{and} \quad \frac{x_1 + x_2 + 5}{x_1} = \frac{18}{8}.$$

Multiplying across gives us two linear equations, which we can easily solve to get $x_1 = 12$ and $x_2 = 10$. So the area of the quadrilateral X is simply $x_1 + x_2 = 22$.

Alternative Solution:

If the reader is familiar with affine transformations, a neat solution can be achieved by applying a shear parallel to BC to make the triangle right-angled at B (noting that such a transformation preserves areas). We can then apply a co-ordinate system such that $A = (0, a)$, $B = (0, 0)$, $C = (2, 0)$, $Q = (0, q)$, $P = (p_x, p_y)$ and $O = (o_x, o_y)$ (the choice of C is arbitrary, but this value simplifies the calculation). Considering the area of $\triangle QBC$, it is easy to see that $q = 18$. Considering $\triangle BOC$ gives $o_y = 10$, and considering $\triangle QOB$ gives $o_x = \frac{8}{9}$. Now, considering the triangles $\triangle POC$ and $\triangle PBC$, which stand on the same base, we can see that $(p_x, p_y) = (\frac{3}{2} o_x, \frac{3}{2} o_y) = (\frac{4}{3}, 15)$. Looking at the x co-ordinates, it is easy to see that P divides AB in the ratio 2:1, so that we can calculate $A = (0, 45)$. So the area of $\triangle ABC$ is 45, and subtracting the three known areas we get once again that the area of X is 22.

8. Evaluate
$$\sum_{n=1}^{\infty} \frac{n+2}{n(n+1)2^n}.$$

Solution:

Let $S_k = \sum_{n=1}^{k} \frac{n+2}{n(n+1)2^n}$. Then the values of S_k for $k = 1, 2, 3, 4, 5$ are

$$\frac{3}{4}, \quad \frac{11}{12}, \quad \frac{31}{32}, \quad \frac{79}{80}, \quad \frac{191}{192}.$$

In each case, we have a fraction of the form $(k-1)/k = 1 - 1/k$. Moreover, motivated by the presence of the 2^n term in the question, we note that for each S_k, 2^k divides the denominator of S_k. In fact, if we divide the first five denominator by 2^k (for the corresponding values of k), we get 2, 3, 4, 5, 6. This leads us to the conjecture that

$$S_k = \sum_{n=1}^{k} \frac{n+2}{n(n+1)2^n} = 1 - \frac{1}{(k+1)2^k}.$$

We can prove this conjecture by induction on k. For $k = 1$, the conjecture is obviously true, so we assume that is it true for $k = m$ for some positive integer m, so that $S_m = 1 - \frac{1}{(m+1)2^m}$. Then we have

$$S_{m+1} = 1 - \frac{1}{(m+1)2^m} + \frac{m+3}{(m+1)(m+2)2^{m+1}} = 1 - \frac{1}{(m+2)2^{m+1}},$$

and so the conjecture holds for $k = m+1$, and hence by induction for all positive integers k.

Now, the original sum is

$$\sum_{n=1}^{\infty} \frac{n+2}{n(n+1)2^n} = \lim_{k \to \infty} \sum_{n=1}^{k} \frac{n+2}{n(n+1)2^n} = 1 - \lim_{k \to \infty} \frac{1}{(k+1)2^k} = 1.$$

9. Show that there do not exist real numbers a, b with

$$\frac{1}{a} + \frac{1}{b} = \frac{1}{a+b}$$

but that there do exist complex numbers with this property.

Solution:

For the given equation to be defined, we must have $a \neq 0$, $b \neq 0$ and $a+b \neq 0$. Under these conditions, we may multiply through by $ab(a+b)$ to get $a(a+b) + b(a+b) = ab$, or $a^2 + ab + b^2 = 0$.

Now if a and b are non-zero reals, then $x = \frac{a}{b}$ must be a real solution to the equation $x^2 + x + 1 = 0$. But since the discriminant of this polynomial is $1^2 - 4.1.1 = -3 < 0$, there are no real solutions, and hence there do not exist real a and b with the given property.

If a and b are allowed to be complex, then we may set $a = 1$, and solve $b^2 + b + 1 = 0$ to get the value of b. Since all polynomials have complex zeros, we know that such a solution must exist.

Although it is not necessary to determine the specific values, we can see that the equation has solutions $b = \omega, \omega^2$ where ω is a complex cube root of unity. In general then, if $a \neq 0$ is a complex number, then $(a, b) = (a, a\omega)$ and $(a, b) = (a, a\omega^2)$ are the solutions of the given equation.

10. Using each of these ten digits 0, 1, 2, 3, 4, 5, 6, 7, 8, 9 once and only once, construct two five digit numbers with the largest possible product.

Solution:

Suppose that we have the two five digit numbers $a = n_5 n_4 n_3 n_2 n_1$ and $b = m_5 m_4 m_3 m_2 m_1$ made up of the given digits whose product is maximal.

If $n_i < n_j$ for any $i > j$, then swapping the digits n_i and n_j will increase the product by $(n_j - n_i)(10^i - 10^j) b > 0$, and similarly for the digits of b, so that the digits of a and b must decrease from left to right.

Now suppose that for some $i > j$, $m_i < n_j$. Then swapping m_i and n_j would increase the product by $(n_j - m_i)(10^i a - 10^j b)$. But since a and b are both five digit numbers, we must have $10^i a > 10^j b$ for $i > j$, and so this would increase the product. So, since the product is maximal, we must have $n_5, m_5 > n_4, m_4 > n_3, m_3 > n_2, m_2 > n_1, m_1$, and so the biggest two digits must be in the first two positions of each number, the next two in the next two positions, and so on.

Without loss of generality, we may assume that $n_5 = 9$, $m_5 = 8$. So we have $a > b$. Now suppose that for some $i \in [1, 4]$ we have $n_i > m_i$. Then swapping these two digits would increase the sum by $(n_i - m_i)(a - b)10^i > 0$. Since the sum is already maximal, we must have $m_i > n_i$ for $i = 1, 2, 3, 4$.

So we have $n_5 > m_5 > m_4 > n_4 > m_3 > n_3 > m_2 > n_2 > m_1 > n_1$, and so the digits are all completely determined. The two numbers are therefore $a = 96420$ and $b = 87531$, whose product is 8439739020.

Fifteenth Superbrain, 1998 — Solutions

1. The ages of Ann and Bill add up to 93. Ann is now three times as old as Bill was at the time when Ann was twice as old as Bill is now. How old is Bill now?

   Solution:

   Let Ann's current age be a and Bill's current age be b. Let t be the amount of time that has passed since Ann was twice as old as Bill is now. Then we can convert the information in the problem into three simultaneous equations:

   $$a + b = 93, \quad a = 3(b - t), \quad a - t = 2b.$$

   Solving these three equations, we get

   $$a = 64\frac{5}{13}, \quad b = 28\frac{8}{13} \text{ and } t = 7\frac{2}{13}.$$

   So Bill is now $28\frac{8}{13}$ years old.

   Remark:

   The solutions to this problem were made deliberately obscure, since problems of this form with whole-number ages are often easily solved by inspection.

2. A man drives a car up a hill half a mile long at 30 miles per hour. How fast must he drive the half mile back down the same hill to average 60 miles per hour for the whole journey?

   Solution:

   This is a trick question, designed to take advantage of the intuitive presumption that all averages must be arithmetic means. It seems natural to assume that driving back down the hill at 90 miles per hour will give the desired average. However, this is not the case.

   Suppose that the length of the journey is x miles. Then the time taken to complete the journey up the hill is $x/30$ hours. If the return journey is made at y miles per hour, then it takes a total of x/y hours, and so the total journey time is $x/30 + x/y$. Obviously the total journey distance is simply $2x$, so we get an average speed for the journey of

   $$\frac{2x}{\frac{x}{30} + \frac{x}{y}} = \frac{60y}{30 + y}$$

miles per hour. Setting this equal to sixty miles per hour, we get $60y/(30 + y) = 60$, or $y = 30 + y$, which has no solution.

In general, for positive y, $y/(y + 30) < 1$, so the average speed is bounded above by 60. In other words, to average 60 miles per hour for the entire journey, the return journey would need to be made at infinite speed.

Remark:

This problem can be solved more immediately, albeit less instructively, by observing that doubling both the length and average speed of a journey leaves the time unchanged, so that the outbound trip takes as long as the full trip, and hence the return journey must be made in zero time.

3. Let $ABCD$ be a plane quadrilateral, all angles of which are less than 180°. Let a, b, c and d be the lengths of the sides and x and y the lengths of the diagonals. Show that

$$2(x + y) > a + b + c + d > x + y.$$

Solution:

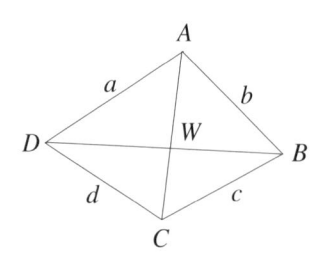

We label the vertices of the figure A, B, C and D (as in the diagram), and label the intersection of the two diagonals W. We suppose that the diagonal of length x is AC, and that the one of length Y is BD. Applying the triangle inequality to each of the four triangles into which the diagonals split the quadrilateral, we get

$|DW| + |WA| > a$, $|AW| + |WB| > b$, $|BW| + |WC| > c$ and $|CW| + |WD| > d$. Adding all four of these together, we get $2((|AW|+|WC|)+ (|BW| + |WD|)) = 2(x + y) > a + b + c + d$, which gives one side of the inequality. To get the other side, we apply the triangle inequality to the four triangles formed by the diagonals and pairs of the sides (i.e. $\triangle ABC, \triangle BCD, \triangle CDA$ and $\triangle DAB$), to get

$$b + c > x, \quad c + d > y, \quad d + a > x \text{ and } a + b > y.$$

Adding the four of these and dividing by two, we get the other side of the inequality: $a + b + c + d > x + y$.

4. Let a, b, c, x, y and z all be integers. Prove or disprove that there exist integers p,q and r with
$$(a^2 + b^2 + c^2)(x^2 + y^2 + z^2) = p^2 + q^2 + r^2.$$

Solution:

It is not hard to find a counter-example to this claim with a little experimentation. The easiest approach is to compile a table of numbers which can be expressed as the sum of three squares of integers, and try to find a small composite value which is not in the list, but the factors of which are.

It is easy enough to find a representation of every integer from 1 to 14 (except 7, which is prime) in the form $x^2 + y^2 + z^2$, but no such representation exists for 15. So, if we set $a = b = c = x = 1$, $y = 2$ and $z = 0$, then we get $(a^2 + b^2 + c^2)(x^2 + y^2 + z^2) = 15 = p^2 + q^2 + r^2$, and there is no solution.

Remark:

It takes quite a bit more work to find a counter example to the claim if a, b, c, x, y, z, p, q and r are positive integers. However, the reader may confirm without too much difficulty that 63 cannot be expressed as a sum of three non-zero squares, but $a = b = c = x = 1$, $y = 2$, $z = 4$ gives $p^2 + q^2 + r^2 = 63$. This is the smallest counter-example.

5. A bag contains 36 snooker balls; some are black and the remainder are white, but there are more black than white. Two balls are drawn from the bag at the same time. If it is equally likely that the balls will be of the same colour as of different colours, how many black balls are in the bag?

Solution:

Suppose that there are b black balls, and hence $36 - b$ white balls. The condition that there be more black balls than white gives $b > 18$.

The number of ways to choose two balls of the same colour is $\binom{b}{2} + \binom{36-b}{2}$ (since we may either choose two black balls or two white balls).

The number of ways to pick two balls of different colours is $b(36 - b)$ (since we may pick any one of the black balls, and any one of the white balls).

Since the number of ways of picking two balls of the same colour is the same as the number of ways of picking two balls of different colours,

we must have
$$\frac{b(b-1)}{2} + \frac{(36-b)(35-b)}{2} = b(36-b).$$

Multiplying through by two and gathering terms to one side, we get $b^2 - 36b + 315 = (b-15)(b-21) = 0$. So we must have $b = 15$ or $b = 21$. But now, we know that $b > 18$, so $b = 21$ is the only solution.

So there must be 21 black balls and 15 white balls in the bag.

6. A circular disc is divided into three pieces, A, B and C of equal area as in the diagram. If $0° < \alpha < 180°$, find an integer n such that $n° < \alpha < (n+1)°$.

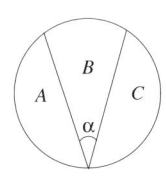

Solution:

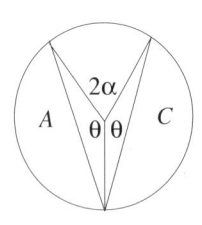

We join the centre of the circle to the two ends of each of the two chords which divide up the circle. Let the angle subtended by each chord be θ (by symmetry, the two must be the same).

But now, the area of a segment of a circle cut off by a chord subtending an angle of θ is known to be

$$A = \frac{1}{2}r^2(\theta - \sin\theta),$$

where r is the radius of the circle. So if the area of each of these segments is to be one third of the area of the circle, we must have $\frac{1}{2}r^2(\theta - \sin\theta) = \frac{1}{3}\pi r^2$, and hence $\theta - \sin\theta = \frac{2}{3}\pi$.

But now, $\frac{d}{d\theta}(\theta - \sin\theta) = 1 - \cos\theta \geq 0$, so $\theta - \sin\theta$ is an increasing function. Since we know that $0 < \theta < \pi$, it is easy to find θ to within an accuracy of one degree by a binary search. We find that $\theta = 149.274\ldots°$.

But now, since the angle subtended at the centre of a circle is twice the angle subtended at the circumference, we get $2\theta + 2\alpha = 2\pi$, and hence $\alpha = \pi - \theta$. So $\alpha = 180° - 149.274\ldots° = 30.725\ldots°$. So we must have $n = 30$.

7. Given a sequence of positive real numbers $a_1, a_2, a_3, \ldots$ with $a_1 = 1$ and
$$\sum_{i=1}^{n} a_i \leq a_{n+1},$$
what is the largest value
$$\sum_{i=1}^{\infty} \frac{1}{a_i}$$
can have?

Solution:

We have $a_1 = 1$, $a_2 \geq a_1 = 1$, $a_3 \geq a_1 + a_2 \geq 2$, $a_4 \geq a_1 + a_2 + a_3 \geq 4$. Continuing in this manner, we get $a_5 \geq 8$, $a_6 \geq 16$, $a_7 \geq 32\ldots$ This leads us to the conjecture that, in general, $a_n \geq 2^{n-2}$, and we may prove this by induction on n.

For $n = 1$, the claim is certainly true. We assume that the conjecture is true for $1 \leq n \leq k$ for some positive integer k, so $a_n \geq 2^{n-2}$ for all $n \in [1, k]$. Then we get

$$a_{k+1} \geq \sum_{i=1}^{k} a_i \geq 1 + \sum_{i=2}^{k} 2^{i-2} = 1 + \sum_{i=0}^{k-2} 2^i = 1 + \left(2^{k-1} - 1\right) = 2^{k-1}.$$

So the conjecture is true for $n = k + 1$, and hence by induction for every positive integer n.

But now, $a_n \geq 2^{n-2}$ gives $1/a_n \leq 2^{2-n}$. Since $\sum_{n=1}^{\infty} 2^{2-n}$ converges absolutely, and since each of the a_i is positive, we have

$$\sum_{n=1}^{\infty} \frac{1}{a_n} \leq 1 + \sum_{n=2}^{\infty} 2^{2-n} = 1 + \sum_{n=0}^{\infty} \frac{1}{2^n} = 3.$$

But now, if we take $a_n = 2^{n-2}$ for $n > 1$, then all of the conditions of the question are met, and the above bound is attained. So the maximum value of $\sum_{n=1}^{\infty} \frac{1}{a_n}$ is simply 3.

8. Evaluate
$$\int \frac{\sin^2 x}{\sin^2 x + 1} dx.$$

Solution:

We may write $\dfrac{\sin^2 x}{\sin^2 x + 1} = 1 - \dfrac{1}{\sin^2 x + 1}$, and hence

$$\int \frac{\sin^2 x}{\sin^2 x + 1}\,dx = x - \int \frac{dx}{\sin^2 x + 1}. \tag{1}$$

Dividing above and below by $\cos^2 x$, we get $\dfrac{1}{\sin^2 x + 1} = \dfrac{\sec^2 x}{\tan^2 x + \sec^2 x}$.
But now, $\sec^2 x = \tan^2 x + 1$, and $\tfrac{d}{dx} \tan x = \sec^2 x$, so we can write

$$\int \frac{dx}{\sin^2 x + 1} = \frac{1}{2}\int \frac{d(\tan x)}{\tan^2 x + \tfrac{1}{2}} = \frac{1}{2}\int \frac{du}{u^2 + a^2} = \frac{1}{2a}\tan^{-1}\frac{u}{a} + C,$$

with $u = \tan x$, $a = \tfrac{1}{\sqrt{2}}$ and C an arbitrary real constant. So, substituting for u and a, and then substituting the entire integral into (1), we get

$$\int \frac{\sin^2 x}{\sin^2 x + 1}\,dx = x - \frac{1}{\sqrt{2}}\tan^{-1}\left(\sqrt{2}\tan x\right) + C.$$

9. Evaluate

$$\sum_{i=1}^{n} 2^i \tan\left(2^i \alpha\right)$$

for values of α for which this sum is defined.

Solution:

We would like to find a telescoping form for the summand — i.e. a function f such that $f(n+1) - f(n) = 2^n \tan(2^n \alpha)$.

Given the form of the difference, we would expect to find a solution of the form $f(n) = 2^n g(2^n \alpha)$, where g is some trigonometric function. Substituting this into the difference equation, we get

$$2^{n+1} g\left(2^{n+1}\alpha\right) - 2^n g\left(2^n \alpha\right) = 2^n \tan\left(2^n \alpha\right).$$

Letting $\beta = 2^n \alpha$, we get $2g(2\beta) - g(\beta) = \tan \beta$. So we wish to find a (most likely trigonometric) function g which satisfies this identity. After some experimentation, we find

$$\cot 2a = \frac{\cos 2a}{\sin 2a} = \frac{\cos^2 a - \sin^2 a}{2 \sin a \cos a} = \frac{\cot a - \tan a}{2},$$

or, after rearranging, $\cot a - 2\cot 2a = \tan a$.

So $g(a) = -\cot a$ satisfies the above identity. Substituting g in the telescoping form above, we get $2^n \cot(2^n \alpha) - 2^{n+1} \cot(2^{n+1} \alpha) = 2^n \tan(2^n \alpha)$. Substituting the telescoping value into the sum, we get

$$\sum_{i=1}^{n} 2^i \tan(2^i \alpha) = 2\cot(2\alpha) - 2^{n+1} \cot(2^{n+1} \alpha)$$

whenever the sum is defined.

10. Solve the equation

$$x^4 + 6x^3 + 11x^2 + 6x - 8 = 0.$$

Solution:

It is easy to confirm using the rational root theorem that this equation has no linear factors with integer coefficients. So, we might attempt to write it as a product of two quadratics.

Suppose $(x^2 + ax + b)(x^2 + cx + d) = x^4 + 6x^3 + 11x^2 + 6x - 8$. Expanding the polynomial on the right hand side and equating coefficients of powers of x, we get four simultaneous equations in a, b, c and d:

$$a + c = 6, \quad b + d + ac = 11, \quad ad + bc = 6 \quad \text{and} \quad bd = -8.$$

If we set $a = c$ in these equations, we get $2a = 6$, and hence $a = 3$. Substituting this into the remaining equations, we get $b + d + 9 = 11$, $3(b+d) = 6$ and $bd = -8$, which can be satisfied by $\{b, d\} = \{-2, 4\}$. So we have $x^4 + 6x^3 + 11x^2 + 6x - 8 = (x^2 + 3x - 2)(x^2 + 3x + 4) = 0$.

But now, we can find the roots of the two quadratics, and hence all of the roots of the original equation. We get

$$x \in \left\{ \frac{-3 + \sqrt{7}i}{2}, \frac{-3 - \sqrt{7}i}{2}, \frac{-3 + \sqrt{17}}{2}, \frac{-3 - \sqrt{17}}{2} \right\}.$$

Remark 1:

We can get a more immediate factorisation by noting the symmetry of the coefficients. To complete this symmetry, we might rewrite the equation as $x^4 + 6x^3 + 11x^2 + 6x + 1 = 9$, and then write the equation as $(x^2 + 3x + 1)^2 = 3^2$. Factoring this as a difference of two squares brings us back to the factorisation obtained above.

Remark 2:

If the inspiration to set $a = c$ to simplify the equations was not forthcoming, it is possible to 'bash it out'. Using the first and fourth equations, we get $c = 6 - a$ and $d = -\frac{8}{b}$. Inserting these into the third equation, we can get a, c, and d in terms of b:

$$a = \frac{6b^2 - b}{b^2 + 8}, \qquad c = \frac{6b + 48}{b^2 + 8}, \qquad d = -\frac{8}{b}.$$

Substituting each of these into the second equation and clearing the denominators, we get a polynomial in b:

$$b^6 - 11b^5 + 44b^4 + 76b^3 - 362b^2 - 704b - 512 = 0.$$

By the rational root theorem, any rational roots of this equation must be of the form $\pm 2^n$ for $n \leq 9$. Trying a few small values, it is not hard to factor this as

$$(b + 2)(b - 4)(b^4 - 9b^3 + 34b^2 + 72b + 64) = 0,$$

and a bit of work confirms that $b = -2$ or $b = 4$ are the only rational roots. We can then insert this back into the equations for a, c, and d and proceed as before.

Sixteenth Superbrain, 1999 — Solutions

1. A man whose birthday was January 1st died on February 29th, 1992. He was x years of age in the year x^2 AD. In what year was he born?

   Solution:

   If the man was x years of age in the year x^2 and died in 1992, then we must have $x^2 < 1992$, and hence $x \leq 44$ (since x is an integer). Also, he must have been $1992 + x - x^2$ when he died.

   But now, if $x \leq 43$, then the man must have been at least $1992 - 43^2 + 43 = 186$ when he died, which is unrealistic. So we must have $x = 44$. So the man was 44 in the year 1936, and was 100 when he died.

2. In the diagram, a square is divided into three regions of equal area A. If $0° < \alpha < 90°$, find α to the nearest degree.

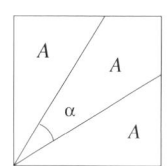

   Solution:

   Consider the right-angled triangle with the angle β (see diagram). The perpendicular sides of this triangle have lengths s and $s \tan \beta$, where s is the side-length of the square, and so it has an area

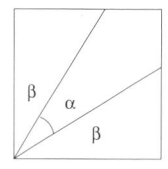

   $A = \frac{1}{2} s^2 \tan \beta = \frac{s^2}{3}$, so we must have $\tan \beta = \frac{2}{3}$. But $\beta \in \left(0, \frac{\pi}{2}\right)$, so we can take inverse tangents to get $\beta = \tan^{-1} \frac{2}{3}$.

   Finally, $2\beta + \alpha = \frac{\pi}{2}$, so $\alpha = \frac{\pi}{2} - 2 \tan^{-1} \frac{2}{3} = 22.619\ldots°$, or 23° to the nearest degree.

3. If p and $p^2 + 14$ are both prime numbers, find all possible values of p.

   Solution:

   The easiest way to approach this problem is to experiment with small values of p, until we have a sense of what is happening.

   For $p = 2$, $2 | (p^2 + 14)$. For $p = 3$, $p^2 + 14 = 23$, which is prime. For $p = 5, 7, 11, 13$ we get $3 | (p^2 + 14)$. So we conjecture that $3 | (p^2 + 14)$ for $p > 3$.

   Since $p = 3$ is the only prime which is divisible by three, any prime $p > 3$ must satisfy $p^2 \equiv 1 \bmod 3$ (since $n^2 \equiv 1 \bmod 3$ whenever $3 \nmid n$). So $p^2 + 14 \equiv 0 \bmod 3$, and hence $3 | (p^2 + 14)$ for all $p > 3$ as conjectured.

But clearly $p^2 + 14 > 3$, so $p^2 + 14$ has a non-trivial factor for $p > 3$ and hence cannot be prime.

So $p = 3$ is the only prime with the property that $p^2 + 14$ is also a prime.

4. If a, b and c are positive integers with $1/a + 1/b + 1/c < 1$, find the maximum possible value of
$$\frac{1}{a} + \frac{1}{b} + \frac{1}{c}.$$

Solution:

We can attempt to maximise the sum $1/a + 1/b + 1/c$ by a 'greedy algorithm' approach (i.e. maximising each term in turn). Since $1/a < 1$, we have $1/a \le 1/2$. So we have $1/b < 1 - 1/2 = 1/2$, and hence $1/b \le 1/3$. Finally,
$$\frac{1}{c} < 1 - \frac{1}{2} - \frac{1}{3} = \frac{1}{6},$$
so $1/c \le 1/7$. So, we have
$$\frac{1}{a} + \frac{1}{b} + \frac{1}{c} \le \frac{41}{42}$$
(based on a greedy algorithm approach). We shall prove that this is the maximum bound.

We may assume that $a \le b \le c$. If we suppose that $a \ge 4$, then we get
$$\frac{1}{a} + \frac{1}{b} + \frac{1}{c} \le \frac{3}{4} < \frac{41}{42},$$
so we must have $a = 2$ or $a = 3$. If $a = 3$, then $b \ge 3$ and $c \ge 4$, so
$$\frac{1}{a} + \frac{1}{b} + \frac{1}{c} \le \frac{11}{12} < \frac{41}{42}.$$
So we must have $a = 2$ in order to maximise the sum.

Now, $b \ge 5$ gives
$$\frac{1}{a} + \frac{1}{b} + \frac{1}{c} \le \frac{9}{10} < \frac{41}{42},$$
and $b = 4$ gives $c \ge 5$, so that
$$\frac{1}{a} + \frac{1}{b} + \frac{1}{c} \le \frac{19}{20} < \frac{41}{42}.$$

So $b \geq 3$, and $b > 3$ means that the sum is not maximal, so we must have $b = 3$. But now, we have $c \geq 7$ as before, so that

$$\frac{1}{a} + \frac{1}{b} + \frac{1}{c} \leq \frac{41}{42}$$

as conjectured.

5. Evaluate

$$\int \frac{\sin\theta + \cos\theta}{3\sin\theta + 2\cos\theta} d\theta.$$

Solution:

If we let $u = 3\sin\theta + 2\cos\theta$, then we get $du = (3\cos\theta - 2\sin\theta)d\theta$, so using this substitution we can integrate

$$\frac{3\cos\theta - 2\sin\theta}{3\sin\theta + 2\cos\theta}$$

with respect to θ.

Also, we can obviously also integrate

$$\frac{3\sin\theta + 2\cos\theta}{3\sin\theta + 2\cos\theta} = 1$$

with respect to θ.

But we can write the integrand as a linear combination of these two quotients if we can find real numbers a and b such that

$$a(3\cos\theta - 2\sin\theta) + b(3\sin\theta + 2\cos\theta) = \sin\theta + \cos\theta.$$

Equating coefficients of $\cos\theta$ and $\sin\theta$ gives $3a + 2b = 1$ and $3b - 2a = 1$. We can solve these simultaneous equations to get $a = 1/13$, $b = 5/13$. So we can write the integral as

$$\int \frac{\sin\theta + \cos\theta}{3\sin\theta + 2\cos\theta} d\theta = \frac{1}{13} \int \frac{d(3\sin\theta + 2\cos\theta)}{3\sin\theta + 2\cos\theta} + \frac{5}{13} \int d\theta$$

$$= \frac{1}{13} \ln|3\sin\theta + 2\cos\theta| + \frac{5\theta}{13} + C,$$

where C is an arbitrary constant.

6. Let ABC be a triangle, and P be a point which lies inside it. Let AP meet BC at X, BP meet CA at Y and CP meet AB at Z. If $\frac{|AP|}{|PX|} = \frac{|BP|}{|PY|} = \frac{|CP|}{|PZ|} = k$, find the value of k.

Solution:

$\triangle ABC$ is partitioned into six smaller triangles by the three Cevians through P. Let these have areas $\alpha_1, \alpha_2, \alpha_3, \alpha_4, \alpha_5$ and α_6 respectively, as in this diagram:

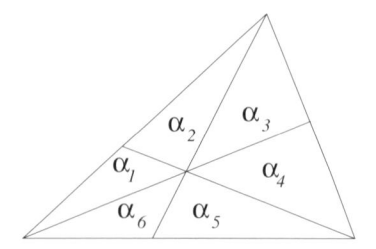

Now, consider the triangles $\triangle CBP$ and $\triangle CPY$. They both have the same altitude (measured from the line BY), so their areas are proportional to their bases. So $\frac{\alpha_5 + \alpha_6}{\alpha_4} = \frac{|BP|}{|PY|} = k$. Similarly, by considering $\triangle BPC$ and $\triangle BZP$, we get $\frac{\alpha_5 + \alpha_6}{\alpha_1} = \frac{|PB|}{|PX|} = k$. So we have $\alpha_1 = \frac{\alpha_5 + \alpha_6}{k} = \alpha_4$.

In a similar manner, we get $\alpha_2 = \alpha_5$ and $\alpha_3 = \alpha_6$.

Now, $\triangle PBZ$ and $\triangle CBZ$ both stand on the same segment of the line AB, and have their apexes at P and C respectively. Similarly, $\triangle PZA$ and $\triangle CZA$ both stand on the same segment of the line AB, and have their apexes at P and C respectively. So, since the area of a triangle standing on a given base is proportional to its height, we get

$$\frac{\alpha_1}{\alpha_1 + \alpha_5 + \alpha_6} = \frac{\alpha_2}{\alpha_2 + \alpha_3 + \alpha_4}.$$

Rearranging, we get $\alpha_1 = \frac{\alpha_1 + \alpha_5 + \alpha_6}{\alpha_2 + \alpha_3 + \alpha_4}\alpha_2 = \frac{\alpha_1 + \alpha_2 + \alpha_3}{\alpha_1 + \alpha_2 + \alpha_3}\alpha_2 = \alpha_2$.

Similarly, $\alpha_3 = \alpha_4$ and $\alpha_5 = \alpha_6$. So combining all of our results, we get

$$\alpha_1 = \alpha_2 = \alpha_3 = \alpha_4 = \alpha_5 = \alpha_6.$$

So, from an earlier equation, we get $k = \frac{\alpha_5 + \alpha_6}{\alpha_4} = \frac{\alpha_1 + \alpha_1}{\alpha_1} = 2$.

Remark:

From the fact that
$$\frac{AP}{PZ} = \frac{BP}{PX} = \frac{CP}{PY} = 2,$$

i.e. that P cuts all three Cevians passing through it in the ratio 2:1, we know that P must be the centroid of the triangle $\triangle ABC$. In fact, this problem proves that the centroid is the only point in any triangle which cuts all three Cevians passing through it in the same ratio.

7. Evaluate
$$\sum_{n=1}^{\infty} \frac{1}{n\sqrt{n+1} + (n+1)\sqrt{n}}.$$

Solution:

We can simplify this problem by clearing the surds from the denominator of the summand, by multiplying above and below by the conjugate. We have

$$\left((n+1)\sqrt{n} + n\sqrt{n+1}\right)\left((n+1)\sqrt{n} - n\sqrt{n+1}\right) = n(n+1),$$

and so we get

$$\frac{1}{n\sqrt{n+1} + (n+1)\sqrt{n}} = \frac{(n+1)\sqrt{n} - n\sqrt{n+1}}{n(n+1)} = \frac{1}{\sqrt{n}} - \frac{1}{\sqrt{n+1}}.$$

So we have constructed a telescoping form for the summand. Substituting this into the sum, we get

$$\sum_{n=1}^{\infty} \frac{1}{n\sqrt{n+1} + (n+1)\sqrt{n}} = \sum_{n=1}^{\infty} \left[\frac{1}{\sqrt{n}} - \frac{1}{\sqrt{n+1}}\right] = 1 - \lim_{k\to\infty} \frac{1}{\sqrt{k}} = 1.$$

8. Show that 128 is not the sum of distinct squares of integers.

Solution:

We note that the sum of the first seven squares is greater than 128, so if 128 can be expressed as a sum of distinct squares, there must be at most six squares in the expression.

Also, we know that if k is any odd integer, then $k^2 \equiv 1 \bmod 8$. So, since 128 is a multiple of eight, the number of odd squares in the expression must be a multiple of eight. But since the number is less than eight, it must be zero. So the expression must contain only squares of even numbers.

Now, if we assume that 128 can be expressed as a sum of squares of distinct even integers, then we may divide across by 4 to get an expression for 32 as a sum of squares of distinct integers. But the sum of the first five squares is greater than 32, so there must be at most four squares in the expression. Also, the sum of the first four squares is 30, and it is easy to see that increasing any of these would increase the sum by more than two, and so an expression as a sum of four squares is also impossible.

Suppose that $32 = a^2 + b^2 + c^2$, where $a > b > c$ are integers. Then $32 > a^2 > 32/3 > 9$, so we must have $a^2 = 16$ or 25, and hence $b^2 + c^2 = 7$ or 16. But it is easy to see that there are no integer solutions to either of these equations.

Finally, suppose that $32 = m^2 + n^2$, where $m > n$ are integers. Then $32 > m^2 > 32/2 = 16$, so we must have $a^2 = 25$, and hence $b^2 = 7$, which has no integer solution.

So 128 cannot be expressed as a sum of distinct squares, as required.

9. If $a_n = x^n/y^n + y^n/x^n$, express $\alpha_5 + \alpha_6$ in terms of α_1 only.

   Solution:

   It is easy to confirm by multiplying out that $\alpha_m \alpha_n = \alpha_{m+n} + \alpha_{m-n}$.
   So $\alpha_2 = \alpha_1^2 - \alpha_0 = \alpha_1^2 - 2$, and $\alpha_4 = \alpha_2^2 - \alpha_0 = (\alpha_1^2 - 2)^2 - 2 = \alpha_1^4 - 4\alpha_1^2 + 2$.
   Then

   $$\alpha_6 = \alpha_2 \alpha_4 - \alpha_2 = (\alpha_1^4 - 4\alpha_1^2 + 2)(\alpha_1^2 - 2) - (\alpha_1^2 - 2) = \alpha_1^6 - 6\alpha_1^4 + 9\alpha_1^2 - 2.$$

   To evaluate α_5, we write

   $$\alpha_5 = \frac{\alpha_4 + \alpha_6}{\alpha_1} = \frac{\alpha_1^6 - 5\alpha_1^4 + 5\alpha_1^2}{\alpha_1} = \alpha_1^5 - 5\alpha_1^3 + 5\alpha_1.$$

   So, we have $\alpha_5 + \alpha_6 = \alpha_1^6 + \alpha_1^5 - 6\alpha_1^4 - 5\alpha_1^3 + 9\alpha_1^2 + 5\alpha_1 - 2$.

   Remark:

   It is easy to see that α_n is expressible as a polynomial in α_1 with integer coefficients for all n. This can be proved by induction using the formula $\alpha_{n+1} = \alpha_n \alpha_1 - \alpha_{n-1}$.

10. A and B are towns 20km and 30km from a straight stretch of river 100km long. Water is pumped from a point P on the river by pipelines to both towns. Where should P be located to minimise the total length of pipe $AP + PB$?

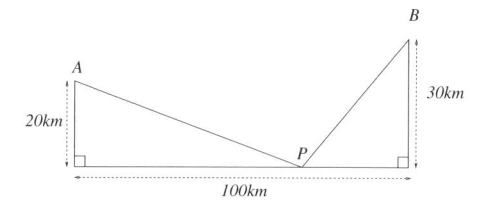

Solution:

Let B' be the reflection of B in the riverbank. Then $PB = PB'$, so that $AP + PB = AP + PB'$. But now, by the triangle inequality, $AP + PB' \geq AB'$, with equality if and only if P is on the straight line joining A to B'.

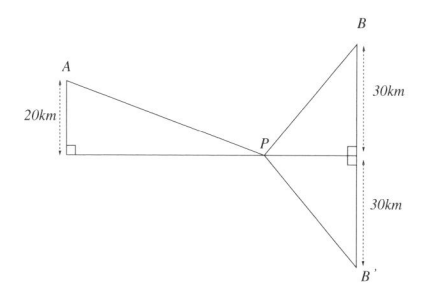

Let the riverbank be the line $y = 0$, $A = (0, 20)$ and $B = (100, -30)$. Then the line AB' has the equation $y = -\frac{1}{2}x + 20$, which intercepts $y = 0$ at $x = 40$. So for $AP+PB$ to be minimal, P should be 40km from the nearest point on the riverbank to A, and 60km from the nearest point on the riverbank to B.

Remark:

Calculations of this form arise naturally in all sorts of reflection problems (which is also the motivation for the method used in the solution) — for example, finding the point on a cushion at which to aim a snooker ball to get from A to B indirectly, finding the point on a mirror to shine a beam of light from A so that it reflects to B, etc.

Seventeenth Superbrain, 2000 — Solutions

1. Four horses A, B, C and D take part in a race. Including dead heats and non-finishers, how many different results are possible?

   Solution:

   We consider the number of results where k horses finish the race for $k = 0, 1, 2, 3, 4$. For $k = 0$, there is obviously only 1 possible result.

   For $k = 1$, the horse that finishes the race wins, and this may be any of A, B, C or D — so there are 4 possible results.

   For $k = 2$, there are six possible pairs of horses which can finish the race. For each pair, either one beats the other, or the two tie, so there are three possible outcomes for each pair, and hence a total of 18 possible results when two horses finish.

   For $k = 3$, there are four possible triples of horses which can finish the race. For each triple of horses, they can finish all in different positions (which gives $3! = 6$ possibilities), one horse can win with the other two tied for second (3 possibilities), one horse can come last with the other two tied for first (3 possibilities), or all three horses can tie (1 possibility). So in total, there are 13 possibilities for each of the 4 triples, and so a total of 52 possible results when all four horses finish the race.

   Finally, we consider the case where all four horses finish. If they all finish in different positions, there are $4! = 24$ different possible results. If two horses tie, they may tie for first, second or third (3 possibilities), there are $\binom{4}{2} = 6$ ways to choose the two, and 2 ways to choose the order of the other two — for a total of 36 possible results. If three horses tie, there are 4 choices for the fourth horse, who can either beat the other three or be beaten by them (2 possibilities), for a total of 8 possible results. If two horses tie for first place and the other two tie for second place, there are $\binom{4}{2} = 6$ possible results. Finally, if all four horse tie, there is one possible result. So if all four horses finish, there are a total of 75 possible results.

   So the total number of possible results is $1+4+18+52+75=150$.

2. If x, y and z are positive integers, prove or disprove that there exist integers p, q and r such that
   $$(x^2 + y^2 + z^2)^2 = p^2 + q^2 + r^2.$$

Solution:

If we expand the left hand side of the equation, we get
$$(x^2 + y^2 + z^2)^2 = x^4 + y^4 + z^4 + 2x^2y^2 + y^2z^2 + 2z^2x^2.$$

Now, if we change the sign of one of the squares, we get
$$(x^2 + y^2 - z^2)^2 = x^4 + y^4 + z^4 + 2x^2y^2 - 2y^2z^2 - 2z^2x^2.$$

So we have $(x^2 + y^2 + z^2)^2 = (x^2 + y^2 - z^2)^2 + 4(x^2 + y^2)z^2$. But now, $4x^2z^2 = (2xz)^2$ and $4y^2z^2 = (2yz)^2$. So if we let $p = x^2 + y^2 - z^2$, $q = 2xz$ and $r = 2yz$, then we get
$$p^2 + q^2 + r^2 = (x^2 + y^2 - z^2)^2 + (2xz)^2 + (2yz)^2 = (x^2 + y^2 + z^2)^2.$$

3. (a) The usual construction for bisecting an angle requires three uses of the compasses:

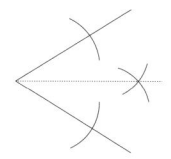

    Show how to bisect and angle with just two uses of the compasses.

   (b) A circle is a set of points all in the plane which a given distance, known as the radius of the circle, from a given point, known as the centre of the circle. Show that a circle has only one centre.

Solution:

(a) The usual construction for bisecting an angle proceeds as follows: First, centre the compasses at the vertex of the angle with an arbitrary radius r and draw a circle. Next, centre the compasses at each of the two points of intersection of this circle with the two branches of the angle in turn with the same radius, and draw a circle. These two circles intersect at the vertex of the angle and at another point on the angle bisector, so joining these two forms the bisector.

We observe that, in this construction, even though only three uses of the compasses are made, one of these uses (the first) intersects the angle twice, and hence creates two points of intersection. By contrast, the other two uses between them create only one further useful point of intersection.

If we were to centre the compasses at the vertex of the angle twice, with a different radius each time, then we would get a total of four points of intersection, symmetrically distributed about the angle bisector. If we then join the intersection of the first compass swing with each branch of the angle to the intersection of the second compass swing with the other branch, then these two lines will intersect on the bisector:

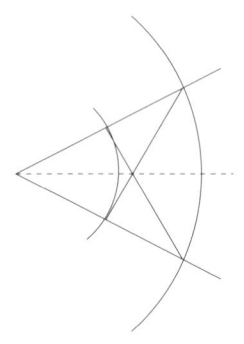

The bisector can then be constructed by joining this point to the vertex of the angle.

(b) Suppose that a circle had two distinct centres. Then there will be a line passing through these two points, which will intersect the circle at two distinct points, say A and B.

But now, each of the centres is on the line AB and equidistant from A and B. But the only such point is the midpoint of AB, and so we cannot have two distinct points with this property. So the circle can have only one centre.

4. An equilateral triangle is dissected into three equal triangles of area A as in the diagram. Find the measure of the angle θ to the nearest degree.

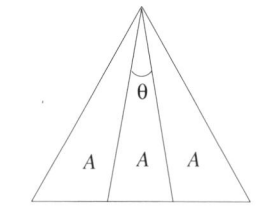

Solution:

Since each of the three triangles has the same area and the same height from the baseline, the baseline must be divided into three equal parts. If we drop a perpendicular from the apex to the base line, then this will bisect the angle θ, and also bisect the baseline.

Let s be the side length of the equilateral triangle. Then its altitude is

$\frac{\sqrt{3}}{2}s$. So we have $\tan\frac{\theta}{2} = \frac{\frac{s}{6}}{\frac{\sqrt{3}}{2}s} = \frac{1}{3\sqrt{3}}$, and hence

$$\tan\theta = \frac{2\tan\frac{\theta}{2}}{1-\tan^2\frac{\theta}{2}} = \frac{\frac{2}{3\sqrt{3}}}{1-\frac{1}{27}} = \frac{3\sqrt{3}}{13}.$$

So we get $\theta = \tan^{-1}\frac{3\sqrt{3}}{13} = 21.786\ldots°$. So, to the nearest degree, the measure of θ is $22°$.

5. Evaluate

$$\int_0^{\frac{\pi}{2}} \frac{\sin^{2000}\theta}{\sin^{2000}\theta + \cos^{2000}\theta}\,d\theta.$$

Solution:

We can solve this integral using the fact that sine and cosine are mutually symmetric on $\left(0, \frac{\pi}{2}\right)$.

If we let $I = \int_0^{\frac{\pi}{2}} \frac{\sin^{2000}\theta}{\cos^{2000}\theta + \sin^{2000}\theta}\,d\theta$, then we can use the rule $\int_a^b f(x)\,dx = \int_a^b f(a+b-x)\,dx$ to get

$$I = \int_0^{\frac{\pi}{2}} \frac{\sin^{2000}\left(\frac{\pi}{2}-\theta\right)}{\cos^{2000}\left(\frac{\pi}{2}-\theta\right) + \sin^{2000}\left(\frac{\pi}{2}-\theta\right)}\,d\theta = \int_0^{\frac{\pi}{2}} \frac{\cos^{2000}\theta}{\cos^{2000}\theta + \sin^{2000}\theta}\,d\theta.$$

Averaging the two expressions for I, we get

$$I = \frac{1}{2}\int_0^{\frac{\pi}{2}} \frac{\cos^{2000}\theta + \sin^{2000}\theta}{\cos^{2000}\theta + \sin^{2000}\theta}\,d\theta = \frac{1}{2}\int_0^{\frac{\pi}{2}} d\theta = \frac{\pi}{4}.$$

Remark:

Although the problem looks more complex, it is essentially the same as question 3 from 1989.

6. Find all positive integers n such that $n^5 + n + 1$ is a prime number.

Solution:

After some experimentation, we find that $n = 1$ is the only value which gives a prime up as far as we may conveniently compute. So we are led to the conjecture that this is the only value of n for which $n^5 + n + 1$

is prime. We attempt to prove this by finding a factorisation of the polynomial.

Now, by the rational root theorem, $n \pm 1$ is the only possible linear factor. Since this is not a factor, we must have a product of a degree two and a degree three polynomial. It is easy to see that each of these must be monic, and have a constant term of ± 1 (where the sign must be the same in both). If we expand $(n^3 + an^2 + bn \pm 1)(n^2 + cn \pm 1)$, and equate coefficients of powers of n with $n^5 + n + 1$, we get $a = -1$, $b = 0$ and $c = 1$, and also we get $-1 \pm 1 = 0$, so that the sign must be plus in both. So we have $n^5 + n + 1 = (n^3 - n^2 + 1)(n^2 + n + 1)$.

For this number to be prime, we must have one of the factors equal to 1, and hence $n^3 - n^2 = 0$ or $n^2 + n = 0$, giving $n \in \{-1, 0, 1\}$. So $n = 1$ is the only positive integer such that $n^5 + n + 1$ is a prime number.

Remark:

We can find the factorisation more quickly by observing that if ω is either of the complex cube roots of unity, then $\omega^5 + \omega + 1 = \omega^2 + \omega + 1 = 0$, so that $(n - \omega_1)(n - \omega_2) = n^2 + n + 1$ must divide $n^5 + n + 1$.

7. If $x = 4t^5 - t^3 + 1$ and $y = 5t^4 + t^2 - t$, find $\frac{d^2y}{dx^2}$ in terms of t only.

Solution:

By the chain rule, $\dfrac{df}{dx} = \dfrac{dt}{dx} \cdot \dfrac{df}{dt} = \left(\dfrac{dx}{dt}\right)^{-1} \dfrac{df}{dt}$, where f is any function of x. So if y is a function of x and x and y are both functions of t, then

$$\frac{d^2y}{dx^2} = \frac{d}{dx}\left(\frac{dy}{dx}\right) = \left(\frac{dx}{dt}\right)^{-1} \frac{d}{dt}\left(\left(\frac{dx}{dt}\right)^{-1} \frac{dy}{dt}\right)$$

Applying the product rule to the derivative on the right hand side gives

$$\frac{d^2y}{dx^2} = \left(\frac{dx}{dt}\right)^{-2} \frac{d^2y}{dt^2} - \left(\frac{dy}{dt}\right)\left(\frac{dx}{dt}\right)^{-3} \frac{d^2x}{dt^2}.$$

Now, it is easy to calculate the first and second derivatives of x and y with respect to t. We find that $\frac{dx}{dt} = 20t^4 - 3t^2$, $\frac{d^2x}{dt^2} = 80t^3 - 6t$, $\frac{dy}{dt} = 20t^3 + 2t + 1$ and $\frac{d^2y}{dt^2} = 60t^2 + 2$. Substituting all of these into the formula for $\frac{d^2y}{dx^2}$, we get

$$\frac{d^2y}{dx^2} = -\frac{400t^5 + 180t^3 + 80t^2 - 6t - 6}{t^5(20t^2 - 3)^3}.$$

8. Factorise $27x^2 - 15xy - 112y^2$.

   Solution:

   This problem is quite straightforward. We can use the quadratic formula to see that $27x^2 - 15xy - 112y^2 = 0$ if
   $$x = \frac{15 \pm \sqrt{15^2 + 4.27.112}}{2.27}y = \frac{15 \pm 111}{54}y = \frac{7}{3}y \text{ or } -\frac{16}{9}y$$
   So we get
   $$27x^2 - 15xy - 112y^2 = (3x - 7y)(9x + 16y).$$

9. Evaluate
   $$\sum_{r=0}^{n} \binom{n}{r}^2.$$

   Solution:

   If we evaluate this sum for $n = 0, 1, 2, 3, 4, \ldots$, we get $1, 2, 6, 20, 70, \ldots$. Now, since there are binomial coefficients in the sum, we should recognise these numbers (it may help to write out Pascal's triangle to a few layers) as the numbers appearing down the centre of Pascal's triangle, i.e. $\binom{0}{0}, \binom{2}{1}, \binom{4}{2}, \binom{6}{3}, \binom{8}{4}, \ldots$. So we are led to the conjecture that
   $$\sum_{r=0}^{n} \binom{n}{r}^2 = \binom{2n}{n}.$$

   To prove this, consider a set A of size $2n$ from which we wish to choose n elements. We partition the set into two subsets, B and C, each of size n. Now, choosing an n-element subset of A is equivalent to choosing an r-element subset of B (with $\binom{n}{r}$ different possible choices) and an $(n-r)$-element subset of C (with $\binom{n}{n-r} = \binom{n}{r}$ different possible choices) for some r with $0 \leq r \leq n$. Summing over all possible values of r, we get $\sum_{r=0}^{n} \binom{n}{r}^2$. But this must be equal to the total number of ways to choose an n-element subset of a set with $2n$ elements, which is $\binom{2n}{n}$.

   Remark 1:

   The solver who is familiar with the identity, $\sum_{r=0}^{p} \binom{n}{r}\binom{m}{p-r} = \binom{m+n}{p}$ (Vandermonde's identity), can solve this problem in a single line by observing that $\sum_{r=0}^{n} \binom{n}{r}^2 = \sum_{r=0}^{n} \binom{n}{r}\binom{n}{n-r} = \binom{2n}{n}$. In fact, the proof given above is essentially similar to the standard combinatorial proof of this identity.

Remark 2:

Both the general Vandermonde identity and this particular case can also be proved by thinking of $\binom{n}{r}$ as binomial coefficients rather than numbers of combinations, and by comparing coefficients of x^p in the equation $(1+x)^m(1+x)^n = (1+x)^{m+n}$ (or, in the particular case, comparing coefficients of x^n in the equation $(1+x)^n(1+x)^n = (1+x)^{2n}$).

10. Four cards, $A1$, $B2$, $C3$ and $D4$ are placed as in the diagram. Place twelve other cards $A2$, $A3$, $A4$, $B1$, $B3$, $B4$, $C1$, $C2$, $C4$, $D1$, $D2$ and $D3$, one in each remaining box, in such a way that no letter or number is repeated in any row, column or (length four) diagonal.

A1	B2	C3	D4

Solution:

This problem can be solved without too much difficulty by a little experimentation, although we shall demonstrate here that a systematic approach is also possible. We approach the problem by separating the letters and the numbers from one another, and solving them as two (almost) independent problems.

First, consider the A and the 1 to go into the second row. To avoid being in a column or a diagonal with $A1$, neither one can be placed in the first two boxes. Also, since there is no other $A1$ card, they cannot go in the same box — so we must have one in the third box and one in the fourth. Since the letters and numbers can essentially be interchanged at this point, we may assume that the 1 goes in the third box, and the A goes in the fourth.

We now have all the information we need to solve the grid of letters and the grid of numbers as independent problems.

There is now only one box in the third row which is not in a column or diagonal with an A (the second), so this must contain the A for that row. There is then only one box in the fourth row (the third) not in a column with A, so this must contain the A for that row.

Now there is only one box (the second) in the last row which is both empty and not in a diagonal or column with the $D4$ card, so this must contain a D. There is then only one box in the second row (the first) not in a column or diagonal with a D, so this must contain the D for that row. Finally, there is then only one box in the third row (the third) not in a column with another D, so this must contain the final D.

Now, each of the two middle columns is missing only one of the four letters, so we may insert a C and a B in the second and third boxes respectively of the second row. Then, each of the two main diagonals is missing only one of the four, so we my insert a C and a B in the first and last boxes respectively of the fourth row. Finally, there is only one letter missing from each of the two extreme columns, and filling these two in completes the grid of letters.

We can approach the grid of numbers in a similar manner. First, we fill in the 1 in the fourth row, and then the third. Then we can fill in the 4 in the fourth, third and second rows (in that order). Then there is only one number missing from each of the second and third columns, and we may fill these in with a 3 and a 2 respectively. Next, each of the two main diagonals is missing exactly one number, so we may insert a 2 and a 3 in the first and fourth boxes respectively of the fourth row. Finally, each of the two extreme columns is missing one number, and filling these in completes the grid.

We can now combine our two solutions to get the answer to the problem:

$A1$	$B2$	$C3$	$D4$
$D3$	$C4$	$B1$	$A2$
$B4$	$A3$	$D2$	$C1$
$C2$	$D1$	$A4$	$B3$

Remark:

From the fact that the construction was determined entirely by the placement of the A and the 1 in the second row, we can conclude that this solution and the solution obtained by swapping the corresponding numbers and letters (*A*s and 1s, *B*s and 2s etc.) above are the only solutions to the problem.

Eighteenth Superbrain, 2001 — Solutions

1. The numbers 123456789 and 246913578 are two examples of nine-digit numbers in which no digit is repeated (we do not include zero as a digit). Find the sum of all such numbers.

   Solution:

   If we select any digit a, there are 8! distinct orderings of the remaining digits, and hence 8! different numbers which have the digit a in the nth position (for $1 \leq a, n \leq 9$). Now, when the digit a is in the nth position (counting from right to left), it contributes $10^{n-1}a$ to the sum. So the total contribution of the digit a (over all positions in all numbers) is

   $$\sum_{n=1}^{9} 8! \times 10^{n-1} a = \left(\frac{10^{10}-1}{9}\right) \times 8!\, a.$$

   Summing this over all the digits, we get the total value of the sum:

   $$\sum_{a=1}^{9} \left(\frac{10^{10}-1}{9} \times 8!\, a\right) = \left(8! \times \frac{10^{10}-1}{9}\right) \sum_{a=1}^{9} a = \frac{10!}{2}\left(\frac{10^{10}-1}{9}\right).$$

   Although this is an acceptable form in which to give the answer, it is not hard to evaluate the sum as a number, giving 2015999999798400.

2. If a, b and c are positive integers which satisfy $a^2 + b^2 = c^2$, show that 60 divides abc.

   Solution:

   To show that $60|abc$, we must show that $3|abc$, $4|abc$ and $5|abc$.

   If $3|a$, $3|b$ or $3|c$, then $3|abc$. So we assume that this is not the case. But now, if some positive integer n is not a multiple of 3, then $n \equiv \pm 1 \mod 3$, and so $n^2 \equiv 1 \mod 3$. So if $3 \nmid abc$, then $a^2 + b^2 \equiv 2 \mod 3$, while $c^2 \equiv 1 \mod 3$, which contradicts the equality of the two. So we must have $3|abc$.

   Now, we know that, for all positive integers n, $n^2 \equiv 0$ or 4 mod 8 if n is even, and $n^2 \equiv 1 \mod 8$ if n is odd. So if a and b are both odd, $c^2 = a^2 + b^2 \equiv 2 \mod 8$, which is impossible. So we may assume that a is even. If either of the other two numbers is even, then we are done, so we assume that neither of them are. Then we get $b^2 \equiv c^2 \equiv 1 \mod 8$ and hence $a^2 \equiv 0 \mod 8$. But if $\frac{a}{2}$ is odd, then $a^2 \equiv 4 \mod 8$, so we must have $\frac{a}{2}$ even. So $4|a$, and hence $4|abc$.

Suppose that none of a, b and c is a multiple of 5. If some positive integer n is not a multiple of 5, then $n \equiv \pm 1$ or $\pm 2 \bmod 5$, and we get $n^2 \equiv \pm 1 \bmod 5$. So we must have $a^2 + b^2 \equiv 0$ or $\pm 2 \bmod 5$ and $c^2 \equiv \pm 1 \bmod 5$, contradicting the equality of the two. To avoid contradiction, we must have $5|abc$.

So we have shown that $3|abc$, $4|abc$ and $5|abc$, and hence $60|abc$, as required.

Remark:

Since $\{a, b, c\} = \{3, 4, 5\}$ gives a case where $abc = 60$, 60 is the largest positive integer which will always divide abc when $a^2 + b^2 = c^2$.

3. Evaluate $\sum_{n=1}^{\infty} \dfrac{n}{n^4 + n^2 + 1}$.

Solution:

We begin this problem by trying to factor the denominator. Since $n^4 + n^2 + 1 \geq 1 > 0$, this polynomial has no real roots, so we seek a factorisation as a product of two quadratics.

If we set $n^4 + n^2 + 1 = (n^2 + an \pm 1)(n^2 + bn \pm 1)$ and equate coefficients of powers of n, we can solve for a, b and the sign of the constant terms. However, we can obtain the answer more neatly by observing that

$$n^4 + n^2 + 1 = n^4 + 2n^2 + 1 - n^2 = (n^2 + 1)^2 - n^2 = (n^2 + n + 1)(n^2 - n + 1).$$

Now, $\dfrac{(n^2 + n + 1) - (n^2 - n + 1)}{2} = n$, and dividing both sides by $n^4 + n^2 + 1$, we get $\dfrac{1}{2(n^2 - n + 1)} - \dfrac{1}{2(n^2 + n + 1)} = \dfrac{n}{n^4 + n^2 + 1}$. But now, $(n-1)^2 + (n-1) + 1 = n^2 - n + 1$, so in telescoping form we have

$$\sum_{n=1}^{\infty} \dfrac{n}{n^4 + n^2 + 1} = \sum_{n=1}^{\infty} \left(\dfrac{1}{2\left((n-1)^2 + (n-1) + 1\right)} - \dfrac{1}{2(n^2 + n + 1)} \right)$$
$$= \lim_{k \to \infty} \left(\dfrac{1}{2(0^2 + 0 + 1)} - \dfrac{1}{2(k^2 + k + 1)} \right) = \dfrac{1}{2}.$$

4. Let AOB be an angle and P a given point. Show how to find a point X on OB and a point Y on OA such that P is the midpoint of XY.

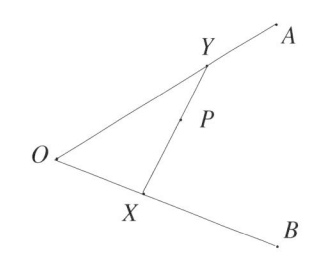

Solution:

The key to this problem is to generalise. Rather than trying to find a single point with a given property, we find all points with the property, and consider the behaviour of the entire set.

In particular, rather than seeking a single point X on the line OB whose reflection in P is a point Y on OA (or equivalently, for which there exists a point Y on OA with P being the midpoint of XY), we seek all points X whose reflection in P is a point Y on OA.

But now, the reflection of the reflection is the point itself, so the set of all points whose reflections lie on the line OA is simply the reflection of the line OA in the point P, which is a line parallel to OA whose distance from P is the same as the distance of OA from P.

Now, since OB intersects OA, it must intersect all lines parallel to OA, including the reflection of OA in P. But this intersection is precisely the point we want, since it lies on OB and its reflection in P lies on OA. So we call this point of intersection X, and let Y be the point of intersection of PX and OA.

5. Express the sixth derivative of the function $y = \tan x$ as a polynomial in $\tan x$.

Solution:

The first derivative of y is $\frac{dy}{dx} = \sec^2 x = \tan^2 x + 1 = y^2 + 1$. But now, by the chain rule, $\frac{df}{dx} = \frac{dy}{dx}\frac{df}{dy}$ for any function f of x.

Letting $f = \frac{dy}{dx}$, we get

$$\frac{d^2y}{dx^2} = \frac{dy}{dx}\frac{d}{dy}\left(\frac{dy}{dx}\right) = (y^2+1)(2y) = 2y^3 + 2y.$$

Letting $f = \frac{d^2y}{dx^2}$, we get

$$\frac{d^3y}{dx^3} = \frac{dy}{dx}\frac{d}{dy}\left(\frac{d^2y}{dx^2}\right) = (y^2+1)(6y^2+2) = 6y^4 + 8y^2 + 2.$$

Continuing in this manner, we get

$$\frac{d^4y}{dx^4} = \frac{dy}{dx}\frac{d}{dy}\left(\frac{d^3y}{dx^3}\right) = 24y^5 + 40y^3 + 16y,$$

$$\frac{d^5y}{dx^5} = \frac{dy}{dx}\frac{d}{dy}\left(\frac{d^4y}{dx^4}\right) = 120y^6 + 240y^4 + 136y^2 + 16,$$

$$\frac{d^6y}{dx^6} = \frac{dy}{dx}\frac{d}{dy}\left(\frac{d^5y}{dx^5}\right) = 720y^7 + 1680y^5 + 1232y^3 + 272y.$$

So we have the sixth derivative of $y = \tan x$ as a polynomial in $\tan x$ (or in y) alone, as required.

6. Evaluate
$$\int \frac{dx}{\sqrt{x} + \sqrt[3]{x}}.$$

Solution:

If we make the substitution $u = \sqrt[6]{x}$, then both $\sqrt{x}$ and $\sqrt[3]{x}$ are powers of u. Moreover, we have $x = u^6$, and hence $dx = 6u^5 du$. Substituting into the integral, we get

$$\int \frac{dx}{\sqrt{x} + \sqrt[3]{x}} = \int \frac{6u^5}{u^3 + u^2} du = \int \left(6u^2 - 6u + 6 - \frac{6}{u+1}\right) du$$

$$= 2u^3 - 3u^2 + 6u - 6\ln(u+1)$$

$$= 2\sqrt{x} - 3\sqrt[3]{x} + 6\sqrt[6]{x} - 6\ln\left(\sqrt[6]{x} + 1\right).$$

Remark:

Both the problem and the approach to the solution are quite similar to problem 5 from 1988.

7. Starting with the first row, place a letter A, B, C, D, E or F in each square so that no letter is repeated in any row, column or length six diagonal.

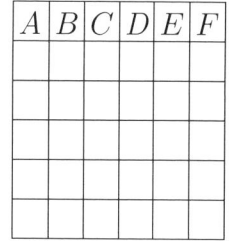

A	B	C	D	E	F

Solution:

There is no easy way to solve this problem completely without a degree of experimentation. However, one easy way to reduce the work in experimenting is to try to find a symmetric solution — that is, a solution in which a reflection in the central axis always maps As to Fs, Bs to Es and Cs to Ds and vice versa. So, whenever we write in an A in the construction, we write in an F in the opposite position, and so on.

The easiest way to approach this problem is almost certainly by filling in the main diagonals first (since then we need only worry about rows and columns). We fill in each letter in turn in its first possible position on the main diagonal — and fill in the off diagonal symmetrically. Then B goes in the third position, since the second is in a column with another B. C goes in the fourth position, since putting a C in the second position would make filling in the remainder of the diagonal impossible (consider the possible locations for the D and the E on the main diagonal). D may go in the second position, and we are then left with F as the only possibility for the fifth position, and hence E as the only possibility for the sixth.

Next, we turn to the second row, and again we insert each letter in its first possible position. A goes in the third position, since the first is in a column with another A. Then B goes in the last position, since the first is in a column with another B.

If we look at the two central columns, there is only one possible entry for each position — E F in the left column and B A in the right. If we then look at the bottom row, the second position must contain a C.

A	B	C	D	E	F
E	D	A	F	C	B
		B	E		
		D	C		
		A	E	B	F
B	C	F	A	D	E

Now if we turn our attention to the fourth column, we see that it is missing an A and a B. Since there is already an B in one of the two empty rows, we may place an A in that row and a B in the other.

Now, the fourth row is missing only an A and an F. But the A cannot go in the first column, since there is already an A there, so it must go in the last column.

We are then left with only the middle two rows of the first and last column to fill, with a C and a D in each row and each column. We may do this either of two ways.
So we have constructed a complete solution.

A	B	C	D	E	F
E	D	A	F	C	B
D	F	B	E	A	C
F	E	D	C	B	A
C	A	E	B	F	D
B	C	F	A	D	E

8. Prove that $\sum_{r=1}^{n} \frac{1}{r^2} < 2$ for each natural number n.

Solution:

There are a number of short solutions to this problem. Firstly, we can make a small adjustment to make the series more like a familiar sum.

In particular, $\sum \frac{1}{r^2}$ is quite like

$$\sum \frac{1}{r(r-1)} = \sum \left(\frac{1}{r-1} - \frac{1}{r}\right),$$

which is a telescoping sum. We wish to find an upper bound for $\sum \frac{1}{r^2}$, so we use the fact that

$$\frac{1}{r^2} < \frac{1}{r(r-1)}$$

for all $r > 1$ to get

$$\sum_{r=1}^{n} \frac{1}{r^2} < 1 + \sum_{r=2}^{n} \frac{1}{r(r-1)} = 1 + \sum_{r=2}^{n} \left(\frac{1}{r-1} - \frac{1}{r}\right) = 2 - \frac{1}{n} < 2.$$

Alternative Solution:

The second approach is in some ways less elegant, but on the other hand is much more easily generalised to similar problems.

Since $f(r) = 1/r^2$ is a decreasing function, we may use an integral comparison to get

$$\sum_{r=1}^{n} \frac{1}{r^2} < 1 + \int_{1}^{n} \frac{dr}{r^2} = 2 - \frac{1}{n} < 2.$$

Remark 1:

It is known that

$$\sum_{r=1}^{\infty} \frac{1}{r^2} = \zeta(2) = \frac{\pi^2}{6} = 1.6449\ldots$$

So, since all the terms are positive, $\sum_{r=1}^{n} \frac{1}{r^2} < \frac{\pi^2}{6} < 1.65$ for all n.

Remark 2:

Note that, in both solutions, it was necessary to separate the first term, and use the respective bounds only on the second and subsequent terms. In each case, this was because the bound for the first term ($1/(1(1-1))$ and $\int_0^1 \frac{dr}{r}$ respectively) was not defined.

9. Find any positive integer values of x, y, z and w which satisfy the equation
$$\frac{1}{x^3} + \frac{1}{y^3} + \frac{1}{z^3} = \frac{1}{w^3}.$$

Solution:

If the reader is familiar with the identity $3^3 + 4^3 + 5^3 = 6^3$ (or manages to find it in the course of exploring of the problem), this may be used to construct a small solution. If we let $x = 3.4 = 12$, $y = 3.5 = 15$ and $z = 4.5 = 20$, then we get

$$\frac{1}{x^3} + \frac{1}{y^3} + \frac{1}{z^3} = \frac{3^3 + 4^3 + 5^3}{3^3 4^3 5^3} = \left(\frac{6}{3.4.5}\right)^3 = \frac{1}{10^3}.$$

Indeed, we might be motivated to seek a solution of the form $x = ab$, $y = bc$, $z = ca$ in any case. We know that if x, y and z have any common factor, then this must also divide w, and to find a small solution, we may cancel this factor. However, if the terms have common factors pairwise then the expression is not so complex when the three fractions are written with a common denominator.

Once the idea has been had to seek a solution of this form, 3, 4, 5 is one of the first possible relatively prime combinations (after 2, 3, 5), and it is easy to find with a little experimentation.

10. The radius r and height h of a right-circular cone are both an integer number of centimetres, and the volume of the cone in cubic centimetres is equal to the total surface area of the cone in square centimetres. Find the values of r and h.

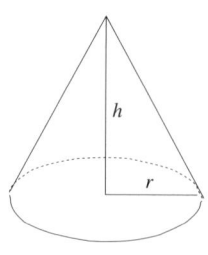

Solution:

Let l be the slant height of the cone. Then $l = \sqrt{r^2 + h^2}$. The total surface area of the figure is $S = \pi r(r+l)$, while its volume is $V = \frac{1}{3}\pi r^2 h$. Setting the two equal to each other and rearranging gives

$$3l = rh - 3r.$$

Squaring both sides, we get $9r^2 + 9h^2 = r^2h^2 - 6r^2h + 9r^2$. Note that $h \neq 0$. Cancelling common terms and then common factors, we get $r^2(h-6) = 9h$. Rearranging, we get $h = \dfrac{6r^2}{r^2-9} = 6 + \dfrac{54}{r^2-9}$. So $\dfrac{54}{r^2-9} = h - 6$ must be an integer, which gives $r^2 \leq 63$. So we must have $r \in \{1, 2, 3, 4, 5, 6, 7\}$.

Testing each case, we see that $(r^2 - 9)|54$ only for $r = 6$, which gives $h = 8$. Substituting these values back into the original equation, we see that they do indeed satisfy the given conditions.

Alternative Solution:

We have $l = \frac{1}{3}rh - r$, and since r and h are integers, l must be rational. But the square root of an integer is either an integer or an irrational number, so l must also be an integer, and hence (r, h, l) is a Pythagorean triple with hypotenuse l.

But it is known that all Pythagorean triples can be expressed in the form $\{a, b, c\} = \{2mnp, p(m^2 - n^2), p(m^2 + n^2)\}$, where c is the hypotenuse, m, n and p are positive integers, $m > n$, and m and n are coprime and of opposite parities.

If $r = 2mnp$ and $h = p(m^2 - n^2)$, then $rh = 3(r + l)$ becomes

$$2mnp^2(m^2 - n^2) = 3p(m^2 + 2mn + n^2).$$

Cancelling common factors, we get $2mnp(m - n) = 3(m + n)$. But $n < m$, so we have $2mnp(m - n) = 3(m + n) < 6m$, and hence $np(m - n) \leq 2$. Since $m - n$ is odd, we must have $\{m, n, p\} = \{2, 1, 1\}$, $\{3, 2, 1\}$ or $\{2, 1, 2\}$, corresponding to $\{r, h, l\} = \{4, 3, 5\}$, $\{12, 5, 13\}$ or $\{8, 6, 10\}$. However, it is easy to confirm that none of these solutions satisfy the original condition.

If $r = p(m^2 - n^2)$ and $h = 2mnp$, then $rh = 3(r + l)$ becomes

$$2mnp^2(m^2 - n^2) = 6pm^2.$$

Cancelling common factors, we get $np(m+n)(m-n) = 3m$. Now, since $n > 0$, we get $np(m - n) = 3\frac{m}{m+n} < 3$, and hence $np(m - n) \leq 2$. This gives rise to the same values of m, n and p as before, and hence to the same triples $\{r, h, l\}$, except that the values of r and h are swapped. Checking the three of these, we find that the only one to satisfy the original condition is $\{r, h\} = \{6, 8\}$.

So once again, we find that $\{r, h\} = \{6, 8\}$ is the only solution.

Nineteenth Superbrain, 2002 — Solutions

1. (a) A leaden sphere of radius 1 metre is melted down and recast as a million spheres of lead shot, each of the same radius. Find the percentage increase in the surface area of the lead exposed.

   (b) To what occupation might questions such as 1(a) be of particular interest?

   Solution:

   (a) Let the radius, surface area and volume of the original sphere be r_0, S_0 and V_0 respectively, and the radius, surface area and volume of each of the recast spheres be r_1, S_1 and V_1 respectively.

   The volume of a sphere is proportional to the cube of its radius (since $V = \frac{4}{3}\pi r^3$), while the exposed surface area is proportional to the square of the radius (since $S = 4\pi r^2$).

   So we have
   $$r_1^3 = \frac{V_1}{V_0} r_0^3 = \frac{r_0^3}{1000000} = \left(\frac{r_0}{100}\right)^3,$$
   and hence $r_1 = r_0/100$. This gives us
   $$S_1 = \frac{r_1^2}{r_0^2} S_0 = \frac{S_0}{10^4}.$$

   So the surface area of each of the recast spheres is one ten thousandth of the surface area of the original sphere, and hence the surface area of all one million spheres must be one hundred times the surface area of the original.

   So the percentage increase must be 9900%.

   (b) This fact would likely be of interest to a chemist (since the rate of many chemical reactions depends on the surface area exposed) among other professions. In particular, this is significant to the storage of grain, which has a massively greater surface area after being milled.

   Some credit was also given on the night of the exam to students who pointed out that 'Superbrain candidates' were another category of people to whom this fact would be of interest.

2. A right-angled triangle has the lengths of its sides in geometric progression. Find r, the common ratio of this geometric progression, correct to four decimal places.

Solution:

Let the sides of the triangle be a, b and c in increasing order. If $r > 1$, then $b = ar$ and $c = ar^2$. Since c is the longest side, it must be the hypotenuse and Pythagoras's theorem gives $a^2 + a^2r^2 = a^2r^4$. So we get $r^4 - r^2 - 1 = 0$ (since $a \neq 0$), and hence $r^2 = (1 \pm \sqrt{5})/2$. Now, $r^2 \geq 0$, so we can only have $r^2 = (1 + \sqrt{5})/2$, and hence

$$r = \sqrt{\frac{1+\sqrt{5}}{2}} \approx 1.2720.$$

If $r < 1$, we must still have $r > 0$, so we mey replace r with $1/p$, and find the value for p as above. So the only other possible value for r must be

$$r = \sqrt{\frac{2}{1+\sqrt{5}}} \approx 0.7862.$$

3. Simplify

$$(1+x)(1+x^2)(1+x^4)(1+x^8)\ldots(1+x^{2^{n-1}}),$$

where x is any real number.

Solution:

Expanding the expression for the first few values of n, we get

$$1+x, \quad 1+x+x^2+x^3, \quad 1+x+x^2+x^3+x^4+x^5+x^6+x^7, \ldots$$

In each case, we have the first 2^n powers of x, so we are led to the conjecture that this is the case in general.

But now, we may use the sum formula for a geometric series to express the sum of the first 2^n powers of x as $(1-x^{2^n})/(1-x)$ (provided that $x \neq 1$, and we will return to the case $x = 1$ later). Our conjecture will be established if we can prove that $\prod_{i=0}^{n-1}(1+x^{2^i}) = \frac{1-x^{2^n}}{1-x}$. We prove this result by induction.

For $n = 1$, we have $(1+x) = (1-x^2)/(1-x)$, and the result is true. Now we assume that the result is true for $n = k$, so

$$\prod_{i=0}^{k-1}\left(1+x^{2^i}\right) = \frac{1-x^{2^k}}{1-x}.$$

We would like to show that the result is true for $n = k+1$.
But now,
$$\prod_{i=0}^{k}\left(1+x^{2^i}\right) = \left(1+x^{2^k}\right)\prod_{i=0}^{k-1}\left(1+x^{2^i}\right),$$
and applying the inductive hypothesis, we get
$$\prod_{i=0}^{k}\left(1+x^{2^i}\right) = \left(1+x^{2^k}\right)\left(\frac{1-x^{2^k}}{1-x}\right) = \frac{1-x^{2^{k+1}}}{1-x}.$$
So the result is established for $n = k+1$, and hence by induction for all positive integers n.

Finally, for the case $x=1$, we get $\left(1+x^{2^k}\right) = 2$ for all positive integers k, so that we get $\prod_{i=0}^{n-1}\left(1+x^{2^i}\right) = 2^n$.

4. A segment of a circular disc has the dimensions shown. Find the area of the segment correct to four decimal places.

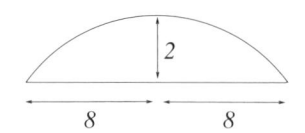

Solution:
Consider the right-angled triangle formed by the centre and one of the end-points of the chord and the centre of the circle. The hypotenuse has length r, while the other two sides have length $r-2$ and 8. So by Pythagoras's Theorem, we have $r^2 = (r-2)^2 + 8^2 = r^2 - 4r + 68$, which gives us $r = 17$.

Now let θ be the angle subtended by the chord. Then $\sin(\theta/2) = 8/17$ and $\cos(\theta/2) = 15/17$. So we get
$$\sin\theta = \frac{2.8.15}{17^2} = \frac{240}{289}.$$

Finally then, the area of the segment is the area of the sector cut off by the radii to the endpoints of the chord less the area of the triangle formed by these two and the chord itself. In other words,
$$A = \frac{1}{2}r^2(\theta - \sin\theta) = \frac{1}{2}(17)^2\left(\sin^{-1}\left(\frac{240}{289}\right) - \frac{240}{289}\right) = 21.5977.$$

172

5. Evaluate
$$\sum_{r=1}^{n} [\tan r\theta \tan(r+1)\theta].$$

Solution:

We would like to write $\tan r\theta \tan(r+1)\theta$ in a telescoping form. In particular, since we are dealing with tan, we would expect a difference of tangents of consecutive multiples of θ, such as $\tan(r+1)\theta - \tan r\theta$. Also, we can have a term of the form $\tan((r+1)-r)\theta = \tan\theta$, since the difference is the same in every case, and can be taken out.

But now, it is possible to express $\tan A \tan B$ as a function of $\tan B - \tan A$ and $\tan(B - A)$ without too much difficulty, by rearranging the familiar rule for the difference of two tangents, to get

$$\tan A \tan B = \frac{\tan B - \tan A}{\tan(B - A)} - 1.$$

So we get

$$\tan r\theta \tan(r+1)\theta = \frac{\tan(r+1)\theta - \tan r\theta}{\tan \theta} - 1.$$

Putting this into the sum gives

$$\sum_{r=1}^{n} [\tan r\theta \tan(r+1)\theta] = \frac{\tan n\theta - \tan\theta}{\tan\theta} - n = \frac{\tan n\theta}{\tan\theta} - n - 1.$$

6. Write the sixth derivative of the function $\sec x$ as a polynomial in $s = \sec x$ only.

Solution:

Let $t = \tan x$. Then $\frac{ds}{dx} = st$, and $\frac{dt}{dx} = s^2$. Also, $t^2 = s^2 - 1$. So we get

$$\frac{d^2 s}{dx^2} = st^2 + s^3 = 2s^3 - s, \quad \frac{d^3 s}{dx^3} = (6s^2 - 1)st = (6s^3 - s)t,$$

$$\frac{d^4 s}{dx^4} = (6s^5 - s^3) + (18s^2 - 1)(st^2) = 24s^5 - 20s^3 + s,$$

$$\frac{d^5 s}{dx^5} = (120s^4 - 60s^2 + 1)(st) = (120s^5 - 60s^3 + s)t,$$

and finally,

$$\frac{d^6 s}{dx^6} = (120s^7 - 60s^5 + s^3) + (600s^4 - 180s^3 + 1)(st^2) = 720s^7 - 840s^5 + 182s^3 - s.$$

7. Prove that at least two different digits occur infinitely often in the decimal expansion of $\pi = 3.141592653589793\ldots$.

Solution:

We assume the contrary, that at most one digit occurs infinitely often in the decimal expansion of π. Since there are infinitely many decimal digits in an expansion, this means that exactly one digit must occur infinitely often.

But now, since all of the other digits together occur only finitely often, there must be a last decimal place in which each of them appears. After this point, the decimal expansion may contain only the one digit which repeats. But this would mean that π has a recurring decimal expansion, and hence is rational. However, it is well known that π is irrational, and hence this cannot be the case.

To avoid contradiction, we must have at least two digits occurring infinitely often in the decimal expansion of π, as required.

Remark:

Since there are ten digits available for each position, it may not seem like a powerful result to say that at least two digits occur in the decimal expansion of π. However, since the method used does not depend on the base in which we are reasoning, this result can equally be applied to the binary expansion of π to show that both 0 and 1 (i.e. all of the binary digits) occur infinitely often in that expansion.

8. Evaluate
$$\int \frac{d\theta}{1+\sin\theta}.$$

Solution 1:

We may write
$$1 + \sin\theta = \sin^2\frac{\theta}{2} + \cos^2\frac{\theta}{2} + 2\sin\frac{\theta}{2}\cos\frac{\theta}{2} = \left(\cos\frac{\theta}{2} + \sin\frac{\theta}{2}\right)^2.$$

Substituting this into the integral and dividing above and below by $\cos^2\frac{\theta}{2}$, we get
$$\int \frac{d\theta}{1+\sin\theta} = \int \frac{\sec^2\frac{\theta}{2}}{\left(1+\tan\frac{\theta}{2}\right)^2} d\theta.$$

Now, if we let $u = 1 + \tan\frac{\theta}{2}$, we get $du = \frac{1}{2}\sec^2\frac{\theta}{2}d\theta$. Substituting this into the integral, we get

$$\int \frac{d\theta}{1+\sin\theta} = 2\int \frac{du}{u^2} = -\frac{2}{u} + C_1 = -\frac{2}{1+\tan\frac{\theta}{2}} + C_1,$$

where C_1 is an arbitrary constant.

Solution 2:

We know that it is possible to write a linear combination of sine and cosine of an angle as the sine of a single angle using the identity

$$a\sin x + b\cos x = \sqrt{a^2+b^2}\sin\left(x + \tan^{-1}\left(\frac{b}{a}\right)\right).$$

Applying this to the identity $1 + \sin\theta = \left(\cos\frac{\theta}{2} + \sin\frac{\theta}{2}\right)^2$, we find that $1 + \sin\theta = 2\sin^2\left(\frac{\theta}{2} + \frac{\pi}{4}\right)$. Substituting this into the integral, we get

$$\int \frac{1}{1+\sin\theta}d\theta = \frac{1}{2}\int \operatorname{cosec}^2\left(\frac{\theta}{2} + \frac{\pi}{4}\right)d\theta.$$

Letting $u = \frac{\theta}{2} + \frac{\pi}{4}$, we get $du = \frac{1}{2}d\theta$, and the integral becomes

$$\int \frac{1}{1+\sin\theta}d\theta = \int \operatorname{cosec}^2 u\, du = -\cot u + C_2 = -\cot\left(\frac{\theta}{2} + \frac{\pi}{4}\right) + C_2,$$

where C_2 is an arbitrary constant.

Solution 3:

Multiplying above and below by $1 - \sin\theta$ in the integrand, we get

$$\int \frac{d\theta}{1+\sin\theta} = \int \frac{1-\sin\theta}{\cos^2\theta}d\theta = \int \sec^2\theta\, d\theta - \int \frac{\sin\theta}{\cos^2\theta}d\theta.$$

Now, $\sec^2\theta$ is the derivative of $\tan\theta$, and the second integral can be evaluated by making a change of variables $u = \cos\theta$. So we get

$$\int \frac{d\theta}{1+\sin\theta} = \tan\theta - \sec\theta + C_3,$$

where C_3 is an arbitrary constant.

Remark:

The first two solutions to this problem serve as a good reminder of the importance of including an arbitrary constant of integration for indefinite integrals. If we were to omit these, we would have

$$-\frac{2}{1+\tan\frac{\theta}{2}} = \int \frac{d\theta}{1+\sin\theta} = -\cot\left(\frac{\theta}{2}+\frac{\pi}{4}\right),$$

which is false (insert any value of θ). However, with the constants included, we get

$$-\frac{2}{1+\tan\frac{\theta}{2}} - \cot\left(\frac{\theta}{2}+\frac{\pi}{4}\right) = C_1 - C_2.$$

Letting $\theta = 0$, we get $C_1 - C_2 = 1$ — so we must have, in general

$$-\frac{2}{1+\tan\frac{\theta}{2}} - \cot\left(\frac{\theta}{2}+\frac{\pi}{4}\right) = 1.$$

This is not hard to confirm by a trigonometric expansion. Similarly, if we drop the constants of integration, the expression for the integral derived in the third solution is equal to that in the second solution, but not to that in the first.

9. Show how to cut a square into four rectangular pieces such that these four pieces can be reassembled to form two squares of different sizes.

Solution:

This question is almost identical to question one from the 1994 superbrain, except for the added requirement that the pieces be rectangular. However, the solution based on dissecting along a uniform grid gives a dissection into rectangles, so the same approach works for this question.

If we dissect a square of side length c into two squares of side length a and b respectively, then $\{a, b, c\}$ is a Pythagorean triple (since we have $a^2 + b^2 = c^2$). Since $\{3, 4, 5\}$ is the smallest integer Pythagorean triple, we might try to find a dissection of a 5×5 square into a 4×4 square and a 3×3 square. It turns out that this is not hard to do:

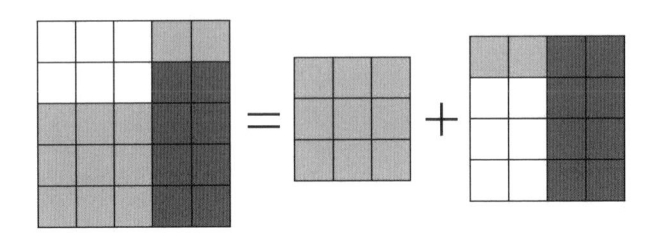

10. Prove that the sum of the squares of five consecutive integers is not the square of an integer.

   Solution:

   Let n be the median of the five numbers, and P be the sum of the five squares. Then $P = (n-2)^2 + (n-1)^2 + n^2 + (n+1)^2 + (n+2)^2 = 5n^2 + 10$. So it is easy to see that $5 | P$.

   But now, if P is a square, then $5 | P$ if and only if $5^2 = 25 | P$. Dividing through by five, we see that this is equivalent to saying that P can be a square if and only if $5 | (n^2 + 2) = P/5$.

   But now, it is easy to verify that all squares of natural numbers are congruent to 0, 1 or 4 modulo 5, and hence $n^2 + 2 \in \{1, 2, 3\} \mod 5$, and $5 \nmid n^2 + 2$. So P cannot be a square, as claimed.

Twentieth Superbrain, 2003 — Solutions

1. A mountaineer is one mile above sea-level. How far away is the furthest point the mountaineer can see on the horizon? (Assume that the Earth is a perfect sphere of radius 3960 miles.)

   Solution:

   Let the mountaineer's location be A, the centre of the earth be B and the furthest point on the horizon that he can see be C. The plane passing through A, B and C passes through the centre of the sphere, and so the intersection of the sphere and the plane is a great circle (i.e. a circle of radius 3960 miles). So, we must have $|AB| = 3961$ miles, $|BC| = 3960$ miles.

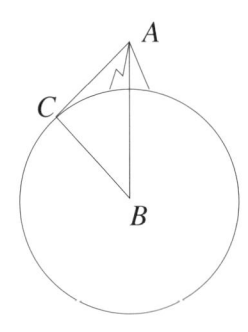

   But now, since AC is a tangent to the circle, it must be perpendicular to BC, and so $\triangle ABC$ is right-angled at C. So we have

   $$|AC| = \sqrt{|AB|^2 - |BC|^2} = \sqrt{3961^2 - 3960^2} \text{ miles} = 89 \text{ miles.}$$

2. The lengths of the sides of a convex quadrilateral are 2, 3, 5 and 6 metres (as in the diagram) and the shorter diagonal has length four metres. Find, correct to four decimal places, the length of the longer diagonal.

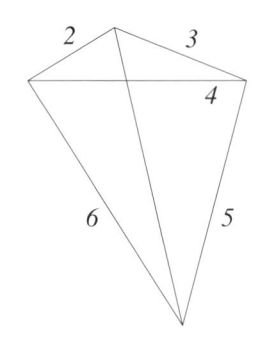

   Solution:

   Let the angle between the side of length 2 and the short diagonal be θ_1, and the angle between the side of length 6 and the short diagonal be θ_2, as in the diagram:

Applying the cosine rule to the triangle of the sides of length 2 and 3 and the short diagonal, we get $\cos\theta_1 = \dfrac{2^2 + 4^2 - 3^2}{2.2.4} = \dfrac{11}{16}$. So we have $\sin\theta_1 = \dfrac{\sqrt{16^2 - 11^2}}{16} = \dfrac{3\sqrt{15}}{16}$. Applying the cosine rule to the triangle of the sides of length 5 and 6 and the short diagonal, we get $\cos\theta_2 = \dfrac{4^2 + 6^2 - 5^2}{2.4.6} = \dfrac{9}{16}$, and so we have $\sin\theta_2 = \dfrac{\sqrt{16^2 - 9^2}}{16} = \dfrac{5\sqrt{7}}{16}$.

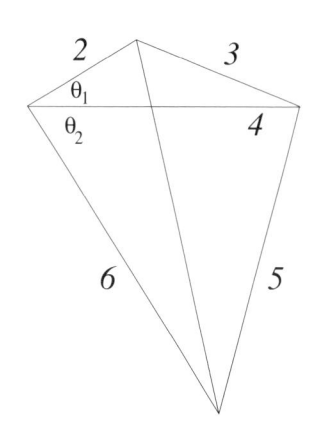

Combining the two, we get

$$\cos(\theta_1 + \theta_2) = \cos\theta_1 \cos\theta_2 - \sin\theta_1 \sin\theta_2 = \dfrac{99 - 15\sqrt{105}}{256}.$$

Finally, we can apply the cosine rule to the triangle of the sides of length 2 and 6 and the long diagonal. If l is the length in metres of the long diagonal, we get

$$l = \sqrt{2^2 + 6^2 - 2.2.6 \left(\dfrac{99 - 15\sqrt{105}}{256} \right)} = 6.7178$$

correct to four decimal places.

3. Find
$$\sum_{r=1}^{n} \dfrac{1}{\sin r\theta \sin(r+1)\theta}.$$

Solution:

We would like to be able to write $\dfrac{1}{\sin r\theta \sin(r+1)\theta}$ in a telescoping form, as $g(r+1, \theta) - g(r, \theta)$. A natural type of function to look for would be one of the form

$$g(r, \theta) = \dfrac{f(r, \theta)}{\sin r\theta},$$

since the difference would then have a denominator of $\sin r\theta \sin(r+1)\theta$ as required. If we rewrite the summand in this form, we get

$$\dfrac{1}{\sin r\theta \sin(r+1)\theta} = \dfrac{f(r+1, \theta)\sin r\theta - f(r, \theta)\sin(r+1)\theta}{\sin r\theta \sin(r+1)\theta}.$$

Now, the numerator looks appealingly similar to the trigonometric expansion $\cos(r+1)\theta \sin r\theta - \cos r\theta \sin(r+1)\theta$, which is simply $-\sin\theta$. Although this is not a constant, it is independent of r, so we can choose

$$f(r,\theta) = -\frac{\cos r\theta}{\sin \theta}$$

to give a numerator of 1 as required.

Substituting this into the telescoping form, we get

$$\sum_{r=1}^{n} \frac{1}{\sin r\theta \sin(r+1)\theta} = \sum_{r=1}^{n} \left[\frac{\cos r\theta}{\sin \theta} - \frac{\cos(r+1)\theta}{\sin \theta}\right] = \frac{\cos \theta - \cos(n+1)\theta}{\sin \theta}.$$

4. If p is a prime number, show that $2p^3 - 1$ and $2p^3 + 1$ cannot both be primes.

   Solution:

   If we calculate $2p^3 - 1$ and $2p^3 + 1$ for the first few primes, and determine the smallest prime factor of each one, we see that 3 divides one of the two values in every case except $p = 3$. So, we conjecture that $3 | 2p^3 \pm 1$ for all prime numbers $p \neq 3$. In order to prove this, we use the properties of primes modulo 3.

   Since $3 \nmid p$ for all primes $p \neq 3$, we must have $p \equiv \pm 1 \bmod 3$. If $p \equiv 1 \bmod 3$, then $2p^3 + 1 \equiv 0 \bmod 3$, and hence $3 | 2p^3 + 1$, and so it cannot be prime. But if $p \equiv -1 \bmod 3$, then $2p^3 - 1 \equiv 0 \bmod 3$, and hence $3 | 2p^3 - 1$, and so it cannot be prime.

   To complete the proof, we observe that in the case $p = 3$, $2p^3 + 1 = 55$, which is not prime. So for all prime numbers p, at least one of $2p^3 - 1$ and $2p^3 + 1$ must be composite.

5. Every point in the plane is coloured red, blue or yellow. Show that there exist two points of the same colour which are a unit distance apart.

   Solution:

   Firstly, we pick any point. We may assume that it is red. Then consider the circle of unit radius centred on this point. If this circle contains any red point, then we have two points of the same colour a unit distance apart — so we assume that this is not the case.

   Now consider a chord on the given circle of unit length. If its endpoints are the same colour, then we are done. So we assume that every chord of unit length on the given circle has endpoints of two different colours. Since there are no red points on the circle, the endpoints must be blue and yellow. But now, for every such chord, there is another point (aside from the first one chosen) which is a unit distance from each of its end points — in fact, it is simply the reflection of the first point in the chord. Now if this point is either blue or yellow, then we have two points of the same colour a unit distance apart, and we are done. So we assume that this is not the case.

So for every chord of unit length on the given circle, the reflection of the centre of the circle in the chord must be red. But now, the locus of the reflected point is simply a circle centred on the original point of radius $\sqrt{3}$. So if we have indeed made the construction so far so as to satisfy the conditions, then we must have an entire red circle, and so we need only take the endpoints of a chord of this circle of unit length to obtain the desired pair.

6. Find the value of
$$\lim_{n \to \infty} \frac{n}{2^n}$$
and justify your answer.

Solution:

Intuitively, it should be easy to see that the exponential function grows much more rapidly than the linear one, and so for sufficiently large n, the expression can be made arbitrarily small. So we expect that $\lim_{n \to \infty} \frac{n}{2^n} = 0$.

To prove this, it is sufficient to show that $2^n > n^2$ for sufficiently large n, since this gives us
$$0 < \frac{n}{2^n} < \frac{n}{n^2} = \frac{1}{n},$$
which tends to zero, and so the result follows by the squeeze (pinching) lemma. This result, that 2^n dominates n^2, is easy to establish by induction.

To prove the base case, we observe that $2^5 = 32 > 25 = 5^2$, so that the conjecture is proved for $n = 5$. Now we assume that the conjecture holds for $n = k$, that is, that $2^k > k^2$. Then $2^{k+1} > 2k^2$. But it is easy to confirm that $2k^2 > (k+1)^2$ for $k > 3$. So, by induction, $2^n > n^2$ for all $n \geq 5$, and the desired result follows as above.

Alternative Solution:

If $a_n = \frac{n}{2^n}$, then $\frac{a_{n+1}}{a_n} = \frac{n+1}{2n} < 1$ for $n > 1$. So a_n is decreasing, but bounded below, since clearly $a_n > 0$ for all n. So a_n must tend to some limit l as n tends to infinity. But then,

$$l = \lim_{n \to \infty} \frac{n+1}{2^{n+1}} = \frac{1}{2} \lim_{n \to \infty} \frac{n}{2^n} + \lim_{n \to \infty} \frac{1}{2^{n+1}} = \frac{1}{2} l + \lim_{n \to \infty} \frac{1}{2^{n+1}}.$$

So we must have $l = \lim_{n \to \infty} \frac{1}{2^n} = 0$.

7. Find all solutions in positive integers of the simultaneous equations

$$x + y = zw, \qquad xy = z + w.$$

Solution:

Firstly, we consider the case where one of the four variables is 1. Since the equations are symmetric in all four variables, we may assume that $z = 1$. So the equations become $x + y = w$ and $xy = w + 1$. Combining the two, we get $xy = x + y + 1$. This expression can be partially factored to give $(x-1)(y-1) = 2$.

Since 2 can be factored as a product of positive integers in only one way, we must have $\{x, y\} = \{2, 3\}$. Substituting this into the equations, we find $w = 5$. Allowing for the various symmetries, this means that we have eight solutions corresponding to the different possible orderings of the variables.

Now we assume that each of the variables in the equation is greater than 1. Suppose, moreover, that at least one of the variables, say x, is greater than two. Then $x \geq 3$, and $xy = z + w \geq 6$. So $zw \geq z(6-z) \geq 8$. So $x + y \geq 8$, and $xy \geq x(8-x) \geq 12$. So $z + w \geq 12$, and $zw \geq z(12-z) = 20$. Continuing in this manner, we get that $xy \geq f^n(6)$ for all n, where $f(x) = 4x - 12$. But f is an increasing, unbounded function, and so xy must be unbounded, which is impossible.

So if each of the variables is greater than 1, then none of them may be greater than 2. However, it is easy to confirm that $x = y = z = w = 2$ is indeed a solution of the given equations.

8. How many rectangular 3×5 tiles can be fitted without overlap in a 17×31 rectangle?

Solution:

A quick calculation confirms that 35 3×5 tiles would cover 525 squares, while $17 \times 31 = 527$. So the answer is at most 35 tiles, and if we can find a packing with 35 tiles, then we are done. It turns out that this is not too hard to do — the key is the 11×11 sub-rectangle which appears in each corner, and which is packed entirely except for one square.

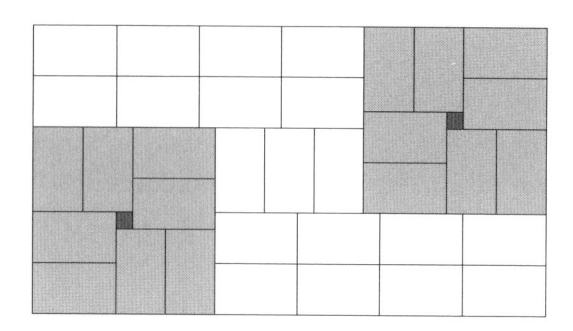

9. The area of a triangle ABC is bisected by a line segment XY (with X on AB and Y on AC) whose length is as small as possible. Show that $|AX| = |AY|$.

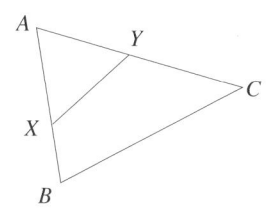

Solution:

The obvious way to approach this problem is to write $|XY|$ as a function of $|AX|$ and $|AY|$, and then seek to minimise it. We can do this easily using the cosine rule:

$$|XY|^2 = |AX|^2 + |AY|^2 - 2|AX||AY|\cos \angle XAY.$$

To minimise this expression, we need to find a relationship between $|AX|$ and $|AY|$. This is where we use the fact that the area of $\triangle AXY$ is exactly half of the area of $\triangle ABC$, which is constant. Let the latter area be Δ. Then we have

$$\frac{1}{2}|AX||AY|\sin \angle XAY = \frac{\Delta}{2}.$$

Rearranging, we get
$$|AY| = \frac{k}{|AX|},$$
where $k = \frac{\Delta}{\sin \angle XAY}$ is a constant. If we let $u = |AX|$ in our expression for $|XY|^2$, then we wish to minimise
$$|XY|^2 = u^2 + \left(\frac{k}{u}\right)^2 - 2k \cos \angle XAY.$$

Differentiating with respect to u, we get $\frac{d}{du}|XY|^2 = 2u - \frac{2k^2}{u^3}$. For $u < \sqrt{k}$, the derivative is negative, and the function is decreasing. For $u > \sqrt{k}$, the derivative is positive, and the function is increasing. So, since the function is continuous, $|XY|^2$ (and hence $|XY|$) must have a global minimum at $u = \sqrt{k}$.

So for $|XY|$ to be minimal, we must have $|AY| = k/\sqrt{k} = \sqrt{k} = |AX|$ as required.

10. The dimensions of an L-shaped figure are given. Locate the centre of gravity (centre of mass, centroid) of the figure. (Assume that the figure has uniform thickness and density.)

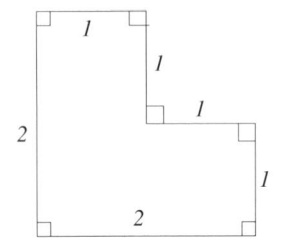

Solution:

We define a system of coordinates, with the origin in the lower left corner of the figure, and the x and y axes extending along the two sides of length 2.

We may consider the given figure as a union of three congruent squares of side 1 unit. But the centre of gravity of a square is simply the centre of the square, so in the given coordinate system the three squares must have their centres at $(\frac{1}{2}, \frac{3}{2}), (\frac{1}{2}, \frac{1}{2})$ and $(\frac{3}{2}, \frac{1}{2})$.

Finally, since each of the squares has the same area, the centre of mass of the entire figure must be located at the point whose coordinates are the unweighted average of the three centres, i.e. at the point $(\frac{5}{6}, \frac{5}{6})$.

Twenty-First Superbrain, 2004 — Solutions

1. In the Irish National Lottery, gamblers are asked to choose six different numbers from the set $\{1, 2, 3, \ldots, 42\}$. What percentage of choices contain at least two consecutive numbers?

   Solution:

   Obviously the total number of possible choices is $\binom{42}{6}$, so we need to determine the total number of choices with at least two consecutive numbers. To do this, it is sufficient to consider the number of choices with no two consecutive numbers.

   But now, if (a, b, c, d, e, f) is an increasing set of six different numbers between 1 and 42 with no two consecutive, then $(a, b-1, c-2, d-3, e-4, f-5)$ is an increasing set of six different numbers between 1 and 37. Similarly, if (a, b, c, d, e, f) is an increasing set of six different numbers between 1 and 37, then $(a, b+1, c+2, d+3, e+4, f+5)$ is an increasing set of six different numbers between 1 and 42 with no two consecutive.

   It is easy to see that these two operations form a bijection between the set of lottery choices with no consecutive numbers and the sets of choices of six different numbers from 1 to 37. So the two sets must be of the same size, and hence the number of lottery choices with no two consecutive numbers is simply $\binom{37}{6}$. So the percentage of choices with no two consecutive numbers is

   $$\binom{37}{6} / \binom{42}{6} = 44.32\%.$$

   So 55.68% of choices contain at least two consecutive numbers.

2. Find all positive integers a and b such that
   $$a^4 + (a+1)^4 + (a+2)^4 = b^4.$$

   Solution:

   It is easy to check that $b^4 \equiv 0$ or $1 \mod 3$. But now, since any three consecutive integers contain each of the three possible values modulo 3 (0, 1 and 2), we must have

   $$b^4 = a^4 + (a+1)^4 + (a+2)^4 \equiv 0^4 + 1^4 + 2^4 \equiv 17 \equiv 2 \mod 3.$$

   So there is no solution to the given equation modulo 3, and hence no solution at all.

3. Prove or disprove that the map of the counties of Ireland can be coloured with three different colours in such a way that counties which touch each other have different colours.

Ireland Counties

Solution:

If we start colouring this map at almost any point by assigning one random county colour 'A' and a random neighbour of it colour 'B', a colouring is then defined by neighbouring sets, and it is easy to arrive at a contradiction showing that a three-colouring is impossible.

The quickest way to arrive at such a contradiction is to start with a county with an odd number of neighbours (Carlow provides the easiest example), and assign that colour 'A'. Then assign one of its neighbours (say Wexford) colour 'B', and proceed clockwise around the first county, so that you must alternate between 'B' and 'C' (i.e. Kilkenny 'C', Laois 'B', Kildare 'C'). But then, since the number of neighbours is odd, the last county arrived at will have a neighbour of each of the three colours, and hence will not be colourable.

4. Evaluate
$$\sum_{n=1}^{\infty} \frac{\sin n\theta}{n!}.$$

Solution:

This sum should remind the solver of the similar series for the exponential function, $e^x = \sum_{n=1}^{\infty} \frac{x^n}{n!}$. However, here we have $\sin n\theta$ rather than $\sin^n \theta$, which would allow us to use this identity.

Considering the problem in this light however, we are reminded of the close relationship between $\sin n\theta$ and $\sin^n \theta$ given by de Moivre's Theorem, $(\cos \theta + i \sin \theta)^n = \cos n\theta + i \sin n\theta$.

Motivated by this, let the given sum be S, and let $C = \sum_{n=1}^{\infty} \frac{\cos n\theta}{n!}$. Then we have

$$C + iS = \sum_{n=1}^{\infty} \frac{\cos n\theta + i\sin n\theta}{n!} = \sum_{n=1}^{\infty} \frac{(\cos\theta + i\sin\theta)^n}{n!} = e^{\cos\theta + i\sin\theta}.$$

So we get $C + iS = e^{\cos\theta} \cos(\sin\theta) + ie^{\cos\theta} \sin(\sin\theta)$. Equating the imaginary parts, we get

$$S = \sum_{n=1}^{\infty} \frac{\sin n\theta}{n!} = e^{\cos\theta} \sin(\sin\theta).$$

5. The area of an equilateral triangle OPQ is bisected by a curve AB of minimal length (where A is on OP and B on OQ). What is the equation of the curve with respect to the given axes?

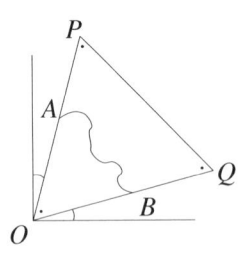

Solution:

Reflecting OPQ repeatedly in OP and OQ and the images of these, we get a hexagon, half of the area of which is enclosed by a closed curve, which must be of minimal length. But now, it is well known that the closed curve of minimal length containing a certain area is the circle (this is known as the isoperimetric property in the plane), and so the curve formed by the repeated reflections of the curve AB must be a circle.

By symmetry, the circle must be centred on the origin, so its equation with respect to the given axes is $x^2 + y^2 = r^2$.

Now let s be the side-length of the given equilateral triangle. Then the area of the hexagon will be

$$(6s^2 \sin 60°)/2 = \frac{3\sqrt{3}}{2}s^2,$$

and so the area of the circle will be $\pi r^2 = \frac{3\sqrt{3}}{4}s^2$. So we have $r^2 = \frac{3\sqrt{3}}{4\pi}s^2$. So the equation of the curve AB with respect to the given coordinate axes is

$$x^2 + y^2 = \frac{3\sqrt{3}}{4\pi}|OP|^2.$$

6. Evaluate

$$\int \frac{dx}{x^4 + 1}.$$

Solution:

Dividing above and below by x^2, we get

$$\int \frac{\frac{1}{x^2}}{x^2 + \frac{1}{x^2}} dx.$$

Now, if we set $v = x + 1/x$ and $u = x - 1/x$, then we can write $1/(x^2 + 1/x^2)$ as $1/(v^2 - 2)$ or as $1/(u^2 + 2)$. Also, we get $dv = (1 - 1/x^2)dx$ and $du = (1 + 1/x^2)dx$. So we can write $(1/x^2)dx = (du - dv)/2$. Substituting these in, we get

$$\int \frac{dx}{x^4 + 1} = \frac{1}{2} \int \frac{du}{u^2 + 2} + \int \frac{dv}{v^2 - 2}.$$

Now, these two are standard integrals, which may be evaluated by setting $u = \sqrt{2} \tan \theta$ and writing

$$\frac{1}{v^2 - 2} = \frac{1}{2\sqrt{2}} \left(\frac{1}{v - \sqrt{2}} - \frac{1}{v + \sqrt{2}} \right).$$

This gives

$$\int \frac{dx}{x^4 + 1} = \frac{1}{2\sqrt{2}} \tan^{-1}\left(\frac{u}{\sqrt{2}}\right) + \frac{1}{4\sqrt{2}} \ln \left| \frac{\sqrt{2} + v}{\sqrt{2} - v} \right| + C$$

$$= \frac{1}{2\sqrt{2}} \tan^{-1}\left(\frac{x - \frac{1}{x}}{\sqrt{2}}\right) + \frac{1}{4\sqrt{2}} \ln \left| \frac{\sqrt{2} + x + \frac{1}{x}}{\sqrt{2} - x - \frac{1}{x}} \right| + C,$$

where C is an arbitrary constant.

7. Find the value of

$$\lim_{n \to \infty} \frac{2^n}{n!},$$

and justify your answer.

Solution:

Evaluating the first few terms, we see that the terms of the sequence are rapidly diminishing, although obviously they are always positive. This suggests that the value of the limit might be zero, but to prove this we need to determine more exactly the rate of decrease.

$\frac{2^n}{n!} = \prod_{r=1}^{n} \frac{2}{r}$, and $\frac{2^4}{4!} < 1$, so that for $n > 4$, we have

$$0 < \frac{2^n}{n!} = \prod_{r=1}^{n} \frac{2}{r} < 1 \cdot \prod_{r=5}^{n} \frac{1}{2} = \frac{1}{2^{n-4}}.$$

But now, $\lim_{n \to \infty} \frac{1}{2^{n-4}} = 0$, so by the squeeze (pinching) lemma, we have $\lim_{n \to \infty} \frac{2^n}{n!} = 0$.

Alternative Solution:

A sufficient condition to show that the limit of a positive sequence is zero is to show that the sum of the corresponding series converges. So, if $\sum_{n=0}^{\infty} \frac{2^n}{n!}$ is finite, then $\lim_{n\to\infty} \frac{2^n}{n!} = 0$. But now, the sum is simply the power series for e^2, and so the result is established.

8. Show how to cut this figure into three pieces and reassemble them to form a square.

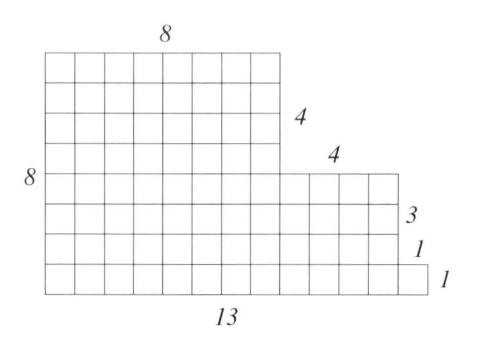

Solution:

It is hard to have a systematic approach to this problem, since it is extremely difficult to prove that certain types of solution are not possible. All that we can do, then, is experiment.

It is easy to see that the square to be formed must be of side length 9. Now, if we remove a strip of length 9 from the bottom right of the figure, then we can dissect the remaining figure into an 8×9 rectangle using a step dissection:

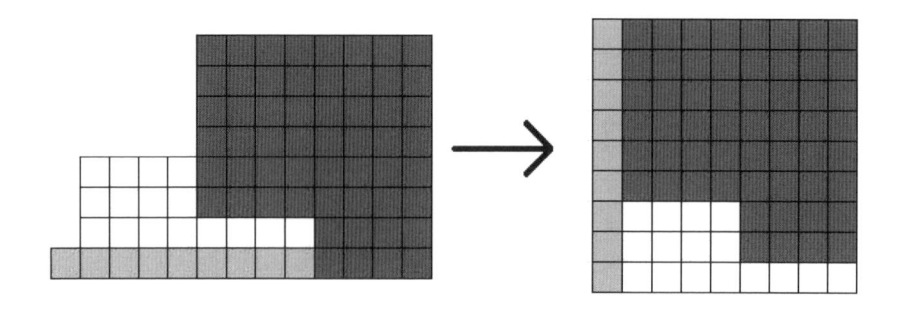

9. Solve the equation $x_1 + x_2 + \ldots + x_n = x_1 x_2 \ldots x_n$ for distinct natural numbers $x_1, x_2, \ldots, x_n$, with $n > 2$.

Solution:

We may assume that the x_i are in increasing order. Since they are all distinct and positive, we must have $x_i \geq i$. So we must have

$$(n-1)! x_n = x_n \prod_{r=1}^{n-1} r \leq \prod_{r=1}^{n} x_n = \sum_{r=1}^{n} x_n < n x_n,$$

and hence $(n-1)! < n$. But now, it is easy to confirm that for $n \geq 4$, $(n-1)! > n$, so we must have $n = 3$.

We must have $x_3 \geq 3$. If $x_3 > 3$, then we have

$$x_1 x_2 x_3 \geq 1.2.x_3 > 3x_3 - 3 \geq x_1 + x_2 + x_3.$$

So we must have $x_3 = 3$, and hence the only solution to the given equation is $(x_1, x_2, x_3) = (1, 2, 3)$.

10. What are the coordinates of the point on the parabola $y^2 = 4x$ which is nearest to the point with coordinates $(-1, 4)$?

Solution:

A point on the parabola $y^2 = 4x$ is of the form $(t^2, 2t)$ for some real number t. The distance from this point to the point $(-1, 4)$ is

$$\sqrt{(t^2 + 1)^2 + (2t - 4)^2} = \sqrt{t^4 + 6t^2 - 16t + 17}.$$

Now, since this distance is positive, and since squaring is an increasing function on the positive reals, the distance will be minimal if and only if its square is minimal — i.e. if and only if $t^4 + 6t^2 - 16t + 17$ is minimal. Let $f(t) = t^4 + 6t^2 - 16t + 17$. Then $f'(t) = 4(t-1)(t^2 + t + 4)$. Now, $t^2 + t + 4 = \left(t + \frac{1}{2}\right)^2 + 3\frac{3}{4} > 0$. So for $t < 1$, $f'(t) < 0$, and for $t > 1$, $f'(t) > 0$. So f decreases on $(-\infty, 1)$ and increases on $(1, \infty)$, and hence has a global minimum at $t = 1$.

So the coordinates of the point on the parabola nearest to the point with coordinates $(-1, 4)$ are $(1^2, 2.1) = (1, 2)$.

Twenty-Second Superbrain, 2005 — Solutions

1. Four equal corners are cut from a square of side 10cm, to leave a regular octagon (i.e. an eight sided figure, all of whose sides have the same length and all of whose angles have the same measure). Find the area of the octagon, correct to three decimal places.

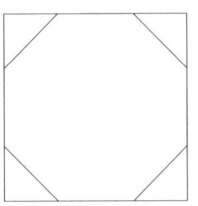

   Solution:

   Let the two equal sides of the triangles cut off each corner have length s cm. Then the sides of the octagon along the sides of the square will have length $10 - 2s$ cm, while the sides at 45° to these will have length $\sqrt{2}s$ cm. Since the octagon is regular, these two must be equal. So we have $10 - 2s = \sqrt{2}s$, or $s = 10/(2 + \sqrt{2}) = 10 - 5\sqrt{2}$.

   Now the area of the octagon is simply the area of the square less four times the area of one of the triangles removed. Denoting this area by A, we get

   $$A = 10^2 - 4\left(\frac{1}{2}\right)(s)(s) = 100 - 2(10 - 5\sqrt{2})^2 = 200(\sqrt{2} - 1) = 82.843.$$

2. Prove that there are infinitely many squares ending in the digits 444.

   Solution:

   The square of an integer can end in 4 only if the integer itself ends in 2 or 8. Using this information to restrict our search, it is not hard to find the first example of a square ending in 444 — namely $38^2 = 1444$. (An even quicker method in this case would be to seek an integer value in the sequence $\sqrt{1000k + 444}$.)

   Having done this, the problem is easily completed, since the last three digits of the square of an integer depend only on the last three digits of the integer itself. So we must have

   $$(1000n + 38)^2 \equiv 444 \bmod 1000$$

   for each integer n, and hence there are infinitely many as required.

Remark:

In fact, it is easy to see that both $(500n + 38)^2$ and $(500n - 38)^2$ both end in 444 for all integers n. Moreover, the last three digits of the square of any integer are entirely determined by its value modulo 500, so it is not hard to confirm that these two cases include all integers whose squares end in 444.

3. Find the value of
$$\sum_{r=0}^{n} \cos(2r+1).$$

Solution:

To express this sum in closed form, we would like to find a telescoping form, and we need a function f such that $f(r+1) - f(r) = \cos(2r+1)$. We can find such a form by using the trigonometric identities relating products and differences of sines and cosines. In particular, if we multiply through the sum by the constant $2\sin(1)$, then we get

$$2\sin(1)\cos(2r+1) = \sin(2r+2) - \sin(2r).$$

In other words, $f(r) = \sin(2r)/(2\sin(1))$ is precisely the function which we need for a telescoping form. Substituting this into the sum, we get

$$\sum_{r=0}^{n} \cos(2r+1) = \sum_{r=0}^{n}\left(\frac{\sin(2r+2)}{2\sin(1)} - \frac{\sin(2r)}{2\sin(1)}\right) = \frac{\sin(2n+2)}{2\sin(1)}.$$

Alternative Solution:

Using the identity $e^{i\theta} = \cos\theta + i\sin\theta$ and the sum of a geometric series, we get

$$\sum_{r=0}^{n} \cos(2r+1) = \operatorname{Re}\left(\sum_{r=0}^{n} e^{i(2r+1)}\right) = \operatorname{Re}\left(\frac{e^i\left(e^{2i(n+1)} - 1\right)}{e^{2i} - 1}\right).$$

But now,
$$\frac{e^i}{e^{2i} - 1} = \frac{1}{2\sinh(i)} = \frac{1}{2i\sin(1)}$$

and
$$e^{2i(n+1)} - 1 = 2e^{i(n+1)}\sinh(n+1) = e^{i(n+1)}i\sin(n+1).$$

Substituting these into the sum, we get

$$\sum_{r=0}^{n} \cos(2r+1) = \operatorname{Re}\left(\frac{e^{i(n+1)}\sin(n+1)}{\sin(1)}\right) = \frac{\cos(n+1)\sin(n+1)}{\sin(1)},$$

which is clearly equivalent to the previous result.

4. Evaluate
$$\int \sqrt{\tan x}\,dx.$$

Solution:

This problem can be solved by taking the 'naive' approach of defining a new variable to be equal to the integrand. If we let $u = \sqrt{\tan x}$, then we get $du = \dfrac{\sec^2 x}{2\sqrt{\tan x}}dx = \dfrac{\tan^2 x + 1}{2\sqrt{\tan x}}dx = \dfrac{u^4+1}{2u}dx$, or $dx = \dfrac{2u}{u^4+1}du$.

Substituting this in gives $\int \sqrt{\tan x}\,dx = \int \dfrac{2u^2}{u^4+1}du$. This integral may be evaluated in a manner similar to problem 6 from 2004.

We begin by dividing above and below by u^2, to get an integrand of $\dfrac{2}{u^2+\frac{1}{u^2}}$. Now, if we let $v = u+1/u$ and $w = u-1/u$, then the integrand may be written as $\dfrac{2}{v^2-2}$ or as $\dfrac{2}{w^2+2}$. Also, $dv = \left(1-\frac{1}{u^2}\right)du$, and $dw = \left(1+\frac{1}{u^2}\right)du$, so that $2du = dv+dw$. Substituting all this into the integral, we get

$$\int \sqrt{\tan x}\,dx = \int \frac{dw}{w^2+2} + \int \frac{dv}{v^2-2}.$$

But now, both of these integrals have well-known solutions. We get

$$\int \sqrt{\tan x}\,dx = \frac{1}{\sqrt{2}}\tan^{-1}\left(\frac{w}{\sqrt{2}}\right) + \frac{1}{2\sqrt{2}}\log\left|\frac{\sqrt{2}-v}{\sqrt{2}+v}\right| + C$$

$$= \frac{1}{\sqrt{2}}\tan^{-1}\left(\frac{\tan x - 1}{\sqrt{2\tan x}}\right)$$
$$+ \frac{1}{2\sqrt{2}}\log\left|\frac{2\sqrt{\tan x}-\tan x-1}{2\sqrt{\tan x}+\tan x+1}\right| + C,$$

where C is an arbitrary constant.

5. Find each of the following:

   (a) $\lim_{x \to \frac{\pi}{2}} [\sec x - \tan x]$.

   (b) $\lim_{n \to \infty} \sqrt[n]{n}$.

Solution:

(a) Clearly we cannot take the two limits separately, since neither sec nor tan has a limit as x approaches $\pi/2$. Writing the limit in terms of sine and cosine gives $(1 - \sin x)/\cos x$. Again, we cannot take the limits of the numerator and the denominator separately, since both tend towards zero, and $\frac{0}{0}$ is undefined. However, we can use this form to simplify the problem somewhat by taking advantage of the relationship between the sine and cosine. To do this, we multiply above and below by $1 + \sin x$, giving:

$$\sec x - \tan x = \frac{1 - \sin x}{\cos x} \cdot \frac{1 + \sin x}{1 + \sin x} = \frac{\cos^2 x}{\cos x (1 + \sin x)} = \frac{\cos x}{1 + \sin x}.$$

But now, we want the limit of a quotient, where the numerator and the denominator are both continuous at the limit, and where the denominator is non-zero at the limit, we may simply insert the limit value to get

$$\lim_{x \to \frac{\pi}{2}} [\sec x - \tan x] = \lim_{x \to \frac{\pi}{2}} \frac{\cos x}{1 + \sin x} = \frac{\cos \frac{\pi}{2}}{1 + \sin \frac{\pi}{2}} = \frac{0}{1+1} = 0.$$

(b) Since the logarithm function is continuous on its domain of definition, and since $\sqrt[n]{n} > 1$ for sufficiently large n (so that the limit falls within this domain), we have

$$\log \left(\lim_{n \to \infty} \sqrt[n]{n} \right) = \lim_{n \to \infty} [\log (\sqrt[n]{n})] = \lim_{n \to \infty} \left[\frac{\log(n)}{n} \right].$$

Now, it is easy to confirm that $n - \log(n) \geq 1$ (note that equality holds at $n = 1$, and that the derivative of the left-hand side is negative for $n < 1$ and positive for $n > 1$), and so that $n > \log n$. Substituting in $\sqrt{n}$ for n, we get $\sqrt{n} > \log(\sqrt{n}) = \frac{1}{2} \log n$. Multiplying through by $\frac{2}{n}$, we get $2/\sqrt{n} > (\log n)/n$. Moreover, it is easy to see that $\frac{\log(n)}{n} > 0$ for $n > 1$. So we have

$$0 = \lim_{n \to \infty} \left[\frac{2}{\sqrt{n}} \right] \geq \lim_{n \to \infty} \left[\frac{\log(n)}{n} \right] \geq 0.$$

Substituting this in above, we get

$$\log\left(\lim_{n\to\infty} \sqrt[n]{n}\right) = \lim_{n\to\infty}\left[\frac{\log(n)}{n}\right] = 0,$$

and hence $\lim_{n\to\infty} \sqrt[n]{n} = e^0 = 1$.

Remark:

It is easy to establish with minimal calculation that the value of the limit in (a) is zero if it exists. The limit can obviously be rewritten as follows:

$$\lim_{x\to 0}\left[\sec\left(\frac{\pi}{2}+x\right) - \tan\left(\frac{\pi}{2}+x\right)\right]$$

But now, each of $\sec\left(\frac{\pi}{2}+x\right)$ and $\tan\left(\frac{\pi}{2}+x\right)$, and hence their difference as well, are odd functions (i.e. functions such that $f(x) = f(-x)$), and it is easy to confirm that if an odd function has a limit as x approaches zero, then that limit must itself be zero.

6. If $AB\|XY\|DC$ and XY contains O, the intersection of AC and BD, find $|XY|$, given that $|AB| = 50$ and $|DC| = 70$.

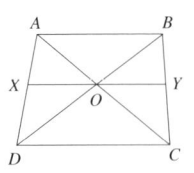

Solution:

Let the perpendicular distance of O from AB and CD respectively be h_1 and h_2. It is easy to see that triangles AOB and COD are similar (consider the alternate angle pairs), and hence that their heights are in proportion to their bases. So, we have $h_2 = 70h_1/50 = 7h_1/5$.

Now, if we extend the sides DA and CB to meet at Z, the triangles ZAB, ZXY and ZDC are all similar (from $AB\|XY\|DC$). So, the ratio of their bases to their heights must be a constant α. If the perpendicular height of Z from AB is h, then we have

$$\alpha = \frac{|AB|}{h} = \frac{|XY|}{h+h_1} = \frac{|CD|}{h+h_1+h_2}.$$

So, we have $20 = |CD| - |AB| = \alpha(h_1 + h_2) = 12h_1\alpha/5$, and hence $h_1\alpha = 25/3$. Finally then,

$$|XY| = \alpha(h + h_1) = |AB| + h_1\alpha = 50 + \frac{25}{3} = \frac{175}{3}.$$

Remark:

If we replace 70 and 50 with a and b, we find that the height of O is $a/(a+b)$, and that the length $|XY|$ becomes $a - a(a-b)/(a+b) = 2ab/(a+b)$, which is the well-known harmonic mean of a and b.

7. A, B and C are workers each of whom works at a steady but different rate. In fact A is the fastest, and C is the slowest. Any of the three working on her own takes a positive integer multiple of the time that the other two, working together, would take on the same job. If C is working on her own, how many times longer does she take on a job than A and B working together?

Solution:

Suppose that A takes a hours to do a certain job, B takes b hours, and C takes c hours. Then A does $1/a$ of the job in one hour, and B does $1/b$ of the job in one hour, so A and B working together do $1/a + 1/b$ of the job in one hour. So A and B working together will take $1/a + 1/b = ab/(a+b)$ hours to do the job. Similarly, B and C working together take $bc/(b+c)$ hours, and C and A working together take $ca/(c+a)$ hours.

Since each one takes an integer multiple of the time taken by the other two, there are integers α, β and γ such that

$$a = \alpha\left(\frac{bc}{b+c}\right), \quad b = \beta\left(\frac{ca}{c+a}\right), \quad c = \gamma\left(\frac{ab}{a+b}\right). \quad (1)$$

Rearranging, we get $\alpha = a/c + a/b$, and similarly for the other two. Multiplying the three equations in (1) together and rearranging, we get

$$\alpha\beta\gamma = \frac{(a+b)(b+c)(c+a)}{abc}$$

$$= \frac{2abc + a^2b + ab^2 + b^2c + bc^2 + c^2a + ca^2}{abc}$$

$$= 2 + \frac{a}{c} + \frac{b}{c} + \frac{b}{a} + \frac{c}{a} + \frac{c}{b} + \frac{a}{b} = \alpha + \beta + \gamma + 2.$$

So we have $\alpha\beta\gamma = \alpha + \beta + \gamma + 2$, where each of α, β and γ are integers. Also, we may derive from the conditions that $\alpha < \beta < \gamma$ (since the work rates of the three are strictly ordered).

If we take $\alpha \geq 2$, then $\beta \geq 3$, so $\alpha\beta\gamma \geq 6\gamma$, while $\alpha + \beta + \gamma + 2 < 4\gamma$, so there can be no solutions. So we must have $\alpha = 1$, from which we get $\beta\gamma = \beta + \gamma + 3$, or $(\beta - 1)(\gamma - 1) = 4$, which has the solution $(\beta, \gamma) = (2, 5)$ ($(\beta, \gamma) = (3, 3)$ is ruled out since $\beta < \gamma$).

So, since C takes γ times longer on a job than A and B together, the desired answer must be $\gamma = 5$.

8. What is the minimum number of pieces into which this figure must be cut so that the pieces can be reassembled to form a square?

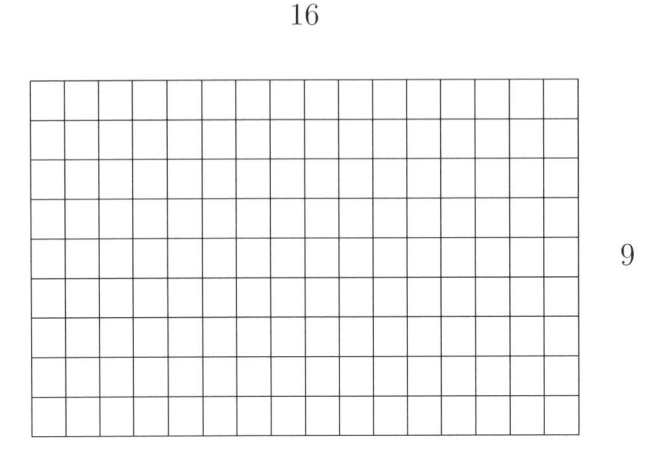

16

9

Solution:

In this problem, it is helpful to take a heuristic approach — rather than merely looking for a solution, we consider what the solution would have to look like. If the minimum number of pieces into which the figure must be cut is n, then we must prove that for all $k < n$, cutting the figure into k pieces is not sufficient. Now, since there is no restriction of any kind made on the cuts which are allowed, such a proof would be very difficult to construct for $k > 1$ (for $k = 1$, the figure is not a square, so the result is obvious) — so it seems highly likely that such a proof is not necessary, which can only be the case if a dissection into two pieces is possible.

Motivated by this fact, we seek a dissection into two pieces which can be reassembled to form a square. In fact, this is relatively easily done using a step dissection:

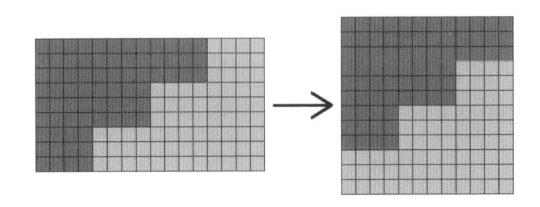

9. For all real numbers x, prove that
$$\sin(\cos x) < \cos(\sin x).$$

Solution:

Obviously we need only consider values of x in any closed interval of length 2π. Moreover, if $x \in [\frac{\pi}{2}, \frac{3\pi}{2}]$, then $\cos x \in [-1, 0]$, and so $\sin(\cos x) < 0$. However, $\sin x \in (-1, 1)$, so $\cos(\sin x) \geq \cos(1) > 0$. So the inequality is true for $x \in [\frac{\pi}{2}, \frac{3\pi}{2}]$, and we need only consider $x \in [-\frac{\pi}{2}, \frac{\pi}{2}]$. Moreover, $\sin(\cos)$ and $\cos(\sin)$ are both even functions, so we need only consider $x \in [0, \frac{\pi}{2}]$.

Now, we note that for $x \in [0, \frac{\pi}{2}]$ (or, indeed, for $x > 0$), we have $\sin x < x$. Also, the cosine function is decreasing on $[0, \frac{\pi}{2}]$. Combining these two facts, we get (for $x \in [0, \frac{\pi}{2}]$) $\sin(\cos x) < \cos x < \cos(\sin x)$, which completes the proof.

10. Evaluate $\displaystyle\prod_{n=2}^{\infty} \frac{n^3 - 1}{n^3 + 1} = \lim_{n \to \infty} \left[\frac{2^3 - 1}{2^3 + 1} \cdot \frac{3^3 - 1}{3^3 + 1} \cdot \frac{4^3 - 1}{4^3 + 1} \cdots \frac{n^3 - 1}{n^3 + 1} \right]$

Solution:

The approach for infinite products is essentially the same as the approach for infinite sums — except that instead of seeking a telescoping sum, we seek a telescoping product. In other words, we would like to express $(n^3 - 1)/(n^3 + 1)$ in the form $f(n)/f(n+1)$.

To find such an expression, we may begin by factoring the given fraction using the well-known identities for the sum and difference of cubes:
$$\frac{n^3 - 1}{n^3 + 1} = \frac{n - 1}{n + 1} \cdot \frac{n^2 + n + 1}{n^2 - n + 1}.$$

We can immediately account for the first term if we let $f_1(n) = n(n-1)$, giving
$$\frac{f_1(n)}{f_1(n+1)} = \frac{n(n-1)}{n(n+1)} = \frac{n-1}{n+1}.$$

So we consider the second term. If we replace n with $n+1$ in the denominator, we get
$$(n+1)^2 - (n+1) + 1 = n^2 + 2n + 1 - n - 1 + 1 = n^2 + n + 1,$$

which is precisely the numerator. So if we set $f_2(n) = 1/(n^2 - n + 1)$, we get
$$\frac{f_2(n)}{f_2(n+1)} = \frac{n^2 + n + 1}{n^2 - n + 1}.$$

Combining the two, we get

$$\frac{n^3-1}{n^3+1} = \frac{f(n)}{f(n+1)},$$

with $f(n) = f_1(n)f_2(n) = (n^2-n)/(n^2-n+1)$.

Using this telescoping form, we may write

$$\prod_{n=2}^{\infty} \frac{n^3-1}{n^3+1} = \lim_{n\to\infty} \frac{f(2)}{f(n+1)} = \lim_{n\to\infty} \frac{2(1-\frac{1}{n})}{3(1-\frac{1}{n}+\frac{1}{n^2})} = \frac{2}{3}.$$

Twenty-Third Superbrain, 2006 — Solutions

1. Find all positive integers n such that $2^n + 1$ is the square of an integer.

   Solution:

   Suppose that $2^n + 1 = k^2$ for some positive integer n and some integer k. Then $2^n = k^2 - 1 = (k-1)(k+1)$. So, there must be non-negative integers a and b with $a + b = n$, $2^a = k - 1$ and $2^b = k + 1$.

   We must have $b > a$, and hence $2^{b-a} = (k+1)/(k-1) = 1 + (2/(k-1))$ must be an integer. So $(k-1)|2$, and hence $k = 2$ or 3. The case $k = 2$ gives $2^n = 3$, which has no positive integer solutions, while $k = 3$ gives $2^n = 8$, which has the unique solution $n = 3$. So $n = 3$ is the only positive integer with the given property.

2. A field is in the shape of a convex quadrilateral (i.e. no angle greater than $180°$). The diagonals of the quadrilateral divide it into four triangles. Three of these triangles have areas 400 square metres, 700 square metres and 800 square metres. What is the largest area the field can have?

   Solution:

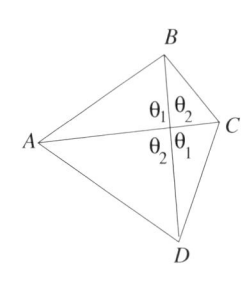

   We label the vertices of the field A, B, C and D, and the point of intersection of the two diagonals O.

   Let $\angle AOB = \theta_1$, $\angle BOC = \theta_2$. Then we also have $\angle COD = \theta_1$ and $\angle DOA = \theta_2$ (opposite angles). Moreover, since $\theta_2 = \pi - \theta_1$, we have $\sin\theta_2 = \sin\theta_1$. Let α, β, γ and δ be the areas of the triangles AOB, BOC, COD and DOA respectively.

   Then we get $2\alpha = \triangle(AOB) = |AO||OB|\sin\theta_1$, $2\beta = 2\triangle(BOC) = |BO||OC|\sin\theta_1$, $2\gamma = 2\triangle(COD) = |CO||OD|\sin\theta_1$ and $2\delta = 2\triangle(DOA) = |DO||OA|\sin\theta_1$. So

   $$\alpha = \frac{\beta\delta}{\gamma} = \frac{\beta\gamma\delta}{\gamma^2}.$$

   So if we let $\{\beta, \gamma, \delta\} = \{400, 700, 800\}$ in some order, then we get that the fourth triangle has an area of $(400.700.800)/a^2$, for $a \in \{400, 700, 800\}$, and so the fourth triangle has an area of at most $(400.700.800)/400^2 = 1400$ square metres, and hence the entire field has an area of no more than $400 + 700 + 800 + 1400 = 3300$ square metres.

3. Evaluate the indefinite integral $\displaystyle\int \frac{dx}{x^n + x}$, where $n > 1$ is a positive integer.

Solution:

The key to this problem is to try to rearrange the denominator into a form that can be more easily worked with. In particular, we would like to be able to make a change of variables to integrate with respect to the denominator. However, if we do this immediately, then $d(x^n + x)$ will contain two terms, when we want only a single term for the numerator.

We may remedy this by dividing above and below by x or x^n, since this will leave only one non-constant term in the denominator. Moreover, we observe that if we divide by x^n, the denominator is of degree $-n+1$, while the numerator is of degree $-n$, which is the same degree as the derivative of the denominator.

So we divide above and below by x^n, and let $u = 1 + x^{-n+1}$, giving $du = -(n-1)x^{-n}dx$. Substituting this into the integral, we get

$$\int \frac{dx}{x^n + x} = \int \frac{x^{-n}}{1 + x^{-n+1}} dx = -\frac{1}{n-1}\int \frac{du}{u} = -\frac{\log|1 + x^{-n+1}|}{n-1} + C,$$

where C is an arbitrary constant.

4. Find all positive integers x and y with $x > y$ such that

$$\frac{1}{x} + \frac{1}{y} = \frac{1}{2006}.$$

Solution:

Multiplying across by $2006xy$, we get $2006x + 2006y = xy$, or, on rearrangement, $(x - 2006)(y - 2006) = 2006^2$. Let $k = y - 2006$. Then $x > y$ gives $2006 > k \geq 1$, and $k | 2006^2$.

Now, $2006^2 = 2^2 \cdot 17^2 \cdot 59^2$, so there are 27 distinct factors of 2006^2 ($2^a 17^b 59^c$, for $a, b, c \in \{0, 1, 2\}$), and hence 13 distinct factors less than 2006. It is easy to find all of these. So we must have

$$k \in \{1, 2, 4, 17, 34, 59, 68, 118, 236, 289, 578, 1003, 1156\}.$$

Now, $\{x, y\} = \left\{\frac{2006^2}{k} + 2006, k + 2006\right\}$, so we can construct all the required solutions:

$$\begin{aligned}\{x, y\} \in \ &\{\{4026042, 2007\}, \{2014024, 2008\}, \{1008015, 2010\}, \\ &\{238714, 2023\}, \{120360, 2040\}, \{70210, 2065\}, \\ &\{61183, 2074\}, \{36108, 2124\}, \{19057, 2242\} \\ &\{15930, 2295\}, \{8968, 2584\}, \{6018, 3009\}, \{5487, 3162\}\}\end{aligned}$$

5. Evaluate
$$\sum_{r=1}^{n}\frac{1}{\sin(2^r x)},$$
where x is a real number and $2^n x$ is not a multiple of π.

Solution:

We may write $1 = \cos^2(2^{r-1}x) + \sin^2(2^{r-1}x) = 2\cos^2(2^{r-1}x) - \cos(2^r x)$. Substituting this into the summand, we get

$$\frac{1}{\sin(2^r x)} = \frac{2\cos^2(2^{r-1}x)}{2\cos(2^{r-1}x)\sin(2^{r-1}x)} - \frac{\cos(2^r x)}{\sin(2^r x)} = \cot(2^{r-1}x) - \cot(2^r x).$$

So we have a telescoping form for the sum (i.e. we may write the summand as $f(r) - f(r-1)$, where $f(r) = -\cot 2^r x$). Putting this into the sum, we get

$$\sum_{r=1}^{n}\frac{1}{\sin(2^r x)} = \sum_{r=1}^{n}\left(\cot(2^{r-1}x) - \cot(2^r x)\right) = \cot x - \cot(2^n x).$$

6. Place the digits 1, 2, 3, 4, 5, 6, 7 and 8, one in each circle, in such a way that no two consecutive digits are placed in circles which are joined directly by a line segment.

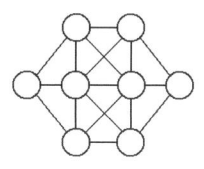

Solution:

Every digit except 1 and 8 is adjacent to two others, and hence can only be linked to $7 - 2 = 5$ of the other digits. But the two central circles each link to six others, and so these can only be filled by 1 and 8. Since the figure is symmetric, we may assume that the 1 goes on the left and the 8 on the right.

Now, 2 must go in the only circle not linking to 1, and similarly 7 must go in the only circle not linking to 8.

Next, 3 must go in one of the two circles not linking to 2. By symmetry, we may assume that it goes in the top left. Similarly, 6 must go in one of the two circle not linking to 7. If it goes on top, then 4 and 5 must fill the two adjacent circles on the bottom, which is impossible. So it must go on the bottom. Finally then, the five must go on top to not be adjacent to the 6, and the 4 must go on the bottom. So we get a final solution like this:

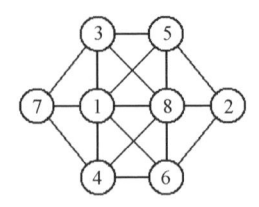

Remark:

It is easy to see by the method used that, up to reflections in the two axes of symmetry, this solution is unique.

7. A water heating tank consists of an open hemisphere (radius r) surmounting an open top right-circular cylinder of radius r and height h. If the total surface area of the tank is a constant A, find the ratio $h : r$ which gives the maximum volume of the tank.

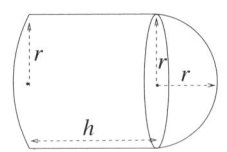

Solution:

Let the volume of the tank be V. Then we have $A = 2\pi rh + 3\pi r^2$ and

$$V = \pi r^2 h + \frac{2}{3}\pi r^3 = \frac{\pi}{3}r^2(2r + 3h).$$

Rearranging the former equation, we get

$$h = \frac{A}{2\pi r} - \frac{3}{2}r.$$

Substituting this value into the expression for V, we find the volume as a function of r alone:

$$V = \frac{\pi}{3}r^2\left(2r + \frac{3A}{2r} - \frac{9}{2}r\right) = \frac{1}{6}\left(3rA - 5\pi r^3\right).$$

But now,

$$\frac{dV}{dr} = \frac{A}{2} - \frac{5\pi r^2}{2}.$$

So for $r \in \left(0, \sqrt{\frac{A}{5\pi}}\right)$, V is decreasing, and for $r \in \left(\sqrt{\frac{A}{5\pi}}, \infty\right)$, V is increasing. So V must have a global minimum with respect to r at

$$r = \sqrt{\frac{A}{5\pi}}.$$

Substituting this value into the expression for h, we find

$$h = \frac{A}{2\pi}\sqrt{\frac{5\pi}{A}} - \frac{3}{2}\sqrt{\frac{A}{5\pi}} = \sqrt{\frac{A}{\pi}}\left(\frac{5}{2\sqrt{5}} - \frac{3}{2\sqrt{5}}\right) = \sqrt{\frac{A}{5\pi}} = r.$$

So, the values of r and h which maximise the volume are equal, and hence the ratio $r : h$ which gives the maximum volume is $1 : 1$.

8. A_1, A_2, A_3, A_4 and A_5 are distinct points in the plane, each of whose coordinates is a pair of integers (i.e. lattice points). Show that there exists a point B and some line segment $A_i A_j$ such that B is a lattice point and B is an internal point of the line segment $A_i A_j$.

Solution:

We consider a stronger version of this problem — in particular, we show that we may assume that the point B is the *midpoint* of the line segment $A_i A_j$ for some integers $1 \leq i < j \leq 5$.

Let $A_i = (x_i, y_i)$. The midpoint of $A_i A_j$ is $\left(\frac{x_i + x_j}{2}, \frac{y_i + y_j}{2}\right)$, which will be a lattice point if and only if $x_i + x_j$ and $y_i + y_j$ are both even, or equivalently if x_i and x_j have the same parity and y_i and y_j have the same parity.

Now, $A_i \equiv (0,0), (0,1), (1,0)$ or $(1,1)$ mod 2. Since there are five of the A_i, and only four different possible pairs of residues modulo two, at least two of the A_i must have the same pair of residues modulo two (by the pigeonhole principle). But now, the midpoint of these two must also be a lattice point, which is precisely the B required.

9. This figure has been cut into four congruent pieces as in the diagram. Show how to cut the figure into four congruent pieces of a different shape to those in the diagram. [Each angle is either $60°$ or $120°$.]

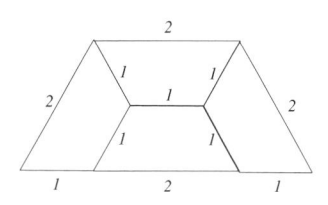

Solution:

The dissection shown in the question cuts the figure along a 'honeycomb' grid (i.e. with lines at angle of 120° to one another). To find an alternative dissection, we try cutting the figure along a square grid.

If we move a vertical line across the figure from each side until they have cut off one quarter of the area each, then we are left with a rectangular area in the middle. Now, if we cut this section in half with a line parallel to one of the slanted sides, then it will be cut into two figures each congruent to the figures cut off on the side, so we are done.

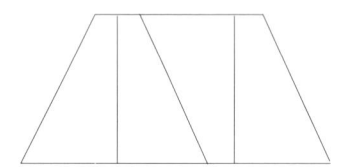

10. The triangular numbers $1, 3, 6, 10, 15, 21, \ldots$ are numbers of the form $n(n+1)/2$ for $n \in \mathbb{N}$. The square numbers $1, 4, 9, 16, 25, \ldots$ are numbers of the form n^2 for $n \in \mathbb{N}$.

    Show that every triangular number greater than 1 is the sum of a square number and two triangular numbers.

    Solution:

    We consider separately the cases where n is even and odd. If n is even, then there is a natural number k such that $n = 2k$. Then

    $$\frac{n(n+1)}{2} = 2k^2 + k = (k^2 + k) + k^2 = \frac{k(k+1)}{2} + \frac{k(k+1)}{2} + k^2.$$

    If $n > 1$ is odd, then there is a natural number k such that $n = 2k+1$. So we can write the nth triangular number as

    $$(2k+1)(k+1) = k(k+1) + (k+1)^2 = \frac{k(k+1)}{2} + \frac{k(k+1)}{2} + (k+1)^2.$$

    Remark:

    In fact, we have established a stronger result, that each triangular number greater than 1 can be expressed as the sum of a square number and twice a triangular number.

Twenty-Fourth Superbrain, 2007 — Solutions

1. Place the numbers 1, 2, 3, 4, 5, 6, 7 and 8 one at each corner of a cube in such a way that the sum of the numbers on each of the six faces of the cube is the same.

   Solution:

   Like with most problems of this kind, the quickest solutions can be achieved mostly by experimentation. However, we shall present a systematic approach to find all solution. We first note that, since any two opposite faces contain each of the corners exactly once between them, the common sum is simply half the sum of all the numbers, which is 18.

   Second, we note that any two intersecting faces have a common sum, and therefore the elements not in common between the two must have a common sum — but this simply means that each pair of edges on opposite sides of the cube has the same sum.

   By using a rotation, we may fix the position of the 1, and label the rest of the cube accordingly. At least one of the three faces containing 1 does not contain 8 (since the only intersection of all three is 1) — suppose that this face is $\{1, b, d, f\}$. Then $b + d + f = 17$, but none of b, d and f is 8, and so we must have $\{b, d, f\} = \{4, 6, 7\}$, and hence $\{a, c, e, g\} = \{2, 3, 5, 8\}$.

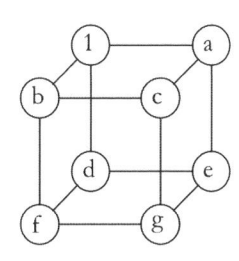

   Now, since opposite edges have equal sums, $a = f + g - 1 \geq 4 + 2 - 1 = 5$, so $a \in \{5, 8\}$. If $a = 5$, then we must have $f = 4$ and $g = 2$. But then, $\{b, d\} = \{6, 7\}$, $\{c, e\} = \{3, 8\}$, and $b - d = e - c$, which clearly has no solution. So we must have $a = 8$.

   But now, $1 + a + b + c = 18$, so $b + c = 9$, $1 + a + d + e = 18$, so $d + e = 9$ and $1 + a = f + g$, so $f + g = 9$. So now, the three edges parallel to $1 - a$ all sum to 9, and the leftmost ends of each are contained in the set $\{b, d, f\} = \{4, 6, 7\}$. It is easy to confirm that any of the six permutations of this set gives rise to a solution, so that there are (up to symmetry) exactly six distinct solutions:

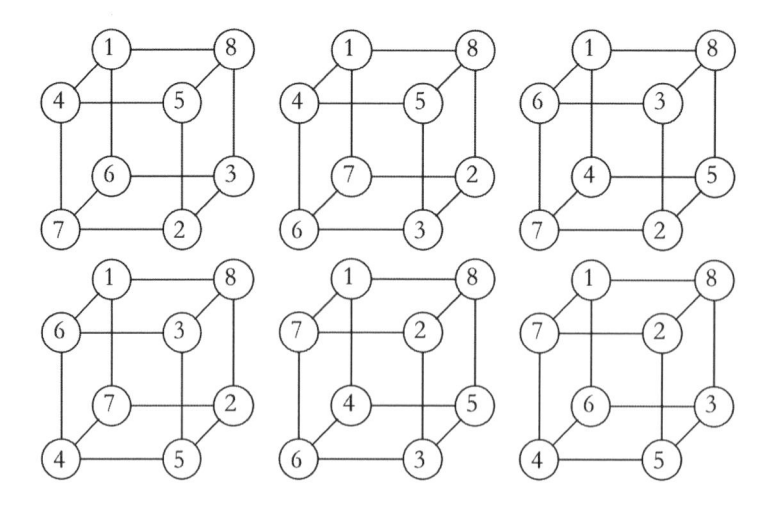

2. The positive integers $8 = 2^3$ and $9 = 3^2$ are consecutive integers each of which is a proper power (ie. of the form a^b where $a, b \in \mathbb{N}$ and $b \geq 2$). Show that there do not exist four consecutive integers each of which is a proper power.

Solution:

Suppose that four such integers exist. Any four consecutive integers must contain one of the form $4n+2$. But now, if $a^b = 4n+2 = 2(2n+1)$, then $2|(4n+2) = a^b$, and hence $2|a$. So we must have $2^b|a^b = 4n+2$, and hence $2^{b-1}|2n+1$.

But now, $b - 1 \geq 1$, but $2n + 1$ is an odd number, so we cannot have $2^{b-1}|2n + 1$. To avoid contradiction, there must not exist four such integers.

Remark:

In fact, it is possible to show that there do not exist two consecutive integers each of which is a perfect power with the exception of the case given in the question ($3^2 - 2^3 = 1$). This remarkable result was conjectured by Eugène Charles Catalan in 1844, and finally settled by Preda Mihăilescu in 2002.

3. You are given six rectangular tiles measuring $1 \times 2, 2 \times 3, 3 \times 4, 4 \times 5, 5 \times 6$ and 6×7 units. What is the area of the smallest integer-sided rectangle into which these tiles can be fitted without overlap? Justify your answer.

Solution:

The sum of the areas of the six tiles is 112, so we know that the rectangle required must have an area of at least 112. If it has an area of exactly

112, then it must have dimensions 7×16 or 8×14 in order to be able to contain the 6×7 tiles (the dimensions must be integers in order for the integer-sided tiles to fit exactly).

If the rectangle is 8×14, then after the 6×7 tile is inserted, it will leave a strip of width 1 or 2 and length at least 6, which cannot be filled with the remaining tiles. If it is 7×16, then the 6×7 tile will leave a similar strip unless it is placed with its 7-unit side parallel to the 7-unit side of the larger rectangle; but if this is done, then the 5×6 strip must be placed in 7×10 rectangle, which must also leave a strip of width 1 or 2 and length at least 5, which cannot be filled. But now, a packing into a 112 square unit rectangle is only possible if the entire area is filled, so in each case this is impossible.

Now, the only way to have a 113 square unit rectangle with integer side-lengths is a 1×113 rectangle, since 113 is prime. Clearly this would not satisfy the requirements, and so the rectangle must be at least 114 square units. In order to contain the 6×7 tile and have integer sidelengths, such a rectangle must have dimensions 6×19. It turns out that it is not hard to achieve a packing of the given tiles into a 6×19 rectangle:

2 X 3	1 X 2	1 X 2		
4 X 3	5 X 4		5 X 6	6 X 7

4. In triangle ABC, X, Y and Z are points on the sides AC, AB and BC respectively such that AZ, BX and CY meet at an internal point O. Prove that
$$\frac{AX}{XC} + \frac{AY}{YB} = \frac{AO}{OZ}.$$

Solution:

Let $\Delta(MNP)$ be the area of the triangle with vertices M, N and P, and let $\Delta_1 = \Delta(AOY)$, $\Delta_2 = \Delta(YOB)$, $\Delta_3 = \Delta(BOZ)$, $\Delta_4 = \Delta(ZOC)$, $\Delta_5 = \Delta(COX)$ and $\Delta_6 = \Delta(XOA)$ (see diagram).

Using the fact that triangles on the same baseline and with the same heights have areas in proportion to their bases, we get
$$\frac{AX}{XC} = \frac{\Delta_6}{\Delta_5} = \frac{\Delta(XAB)}{\Delta(XCB)} = \frac{\Delta_1 + \Delta_2 + \Delta_6}{\Delta_3 + \Delta_4 + \Delta_5}.$$

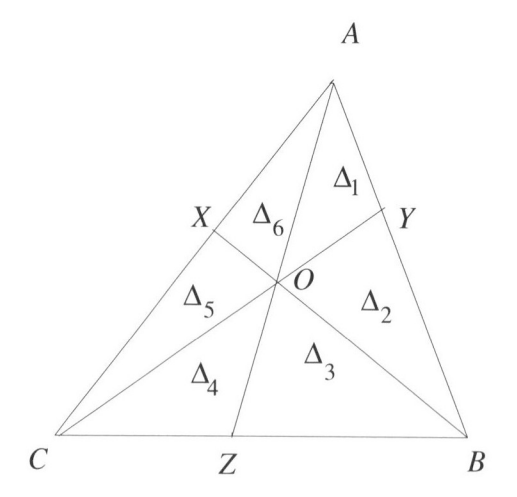

But now, given positive numbers a, b, c and d, we have

$$\frac{a}{b} = \frac{c}{d} \text{ if, and only if, } \frac{a}{b} = \frac{a+c}{b+d}.$$

It is easy to check this by multiplying through and cancelling common terms. So we must have

$$\frac{AX}{XC} = \frac{\Delta_6}{\Delta_5} = \frac{(\Delta_1 + \Delta_2) + \Delta_6}{(\Delta_3 + \Delta_4) + \Delta_5} = \frac{\Delta_1 + \Delta_2}{\Delta_3 + \Delta_4}.$$

By a similar process, we get

$$\frac{AY}{YB} = \frac{\Delta_1}{\Delta_2} = \frac{\Delta_1 + (\Delta_5 + \Delta_6)}{\Delta_2 + (\Delta_3 + \Delta_4)} = \frac{\Delta_5 + \Delta_6}{\Delta_3 + \Delta_4}.$$

A similar method again (but this time using the 'only if' part of the above result) gives

$$\frac{AO}{OZ} = \frac{\Delta_5 + \Delta_6}{\Delta_4} = \frac{\Delta_1 + \Delta_2}{\Delta_3} = \frac{(\Delta_1 + \Delta_2) + (\Delta_5 + \Delta_6)}{\Delta_3 + \Delta_4} = \frac{AX}{XC} + \frac{AY}{YB}.$$

5. If $x + \frac{1}{x} = 2\cos\theta$, show that $x^n + \frac{1}{x^n} = 2\cos n\theta$, where n is an integer.

Solution:

Multiplying the given equation through by x and gathering terms to one side, we get a quadratic equation which can be easily solved:

$$x^2 - 2x\cos\theta + 1 = 0 \quad \Rightarrow \quad x = \cos\theta \pm \sqrt{\cos^2\theta - 1}.$$

Now, $\cos^2\theta - 1 = -\sin^2\theta$, so $\sqrt{\cos^2\theta - 1} = i|\sin\theta|$. So the expression for x simplifies to

$$x = \cos\theta \pm i\sin\theta = e^{\pm i\theta}.$$

Note that the absolute value sign can now be dropped, since the only effect it could have would be to swap the two solutions. Now, we can calculate $x^n + x^{-n}$ easily:

$$x^n + \frac{1}{x^n} = e^{\pm in\theta} + e^{\mp in\theta} = e^{in\theta} + e^{-in\theta}.$$

Note that the $\pm$ and $\mp$ could be dropped, since in either case there is one plus and one minus in the sum. So we have

$$x^n + \frac{1}{x^n} = (\cos n\theta + i \sin n\theta) + (\cos n\theta - i \sin n\theta) = 2\cos n\theta$$

as required

6. Find any set of positive integers $\{a, b, c, d\}$ such that $a^4 + b^5 + c^6 = d^7$.

   Solution:

   There is a simple way to generate solutions to this question by taking advantage of the identity $2^n + 2^n = 2^{n+1}$.

   First, let $\{a, b, c, d\} = \{2^\alpha, 2^\beta, 2^\gamma, 2^\delta\}$. Then we have $2^{4\alpha} + 2^{5\beta} + 2^{6\gamma} = 2^{7\delta}$. If we can pick α, β, γ and δ to turn this into $2^{n-1} + 2^{n-1} + 2^n = 2^{n+1}$ for some n, then we have found a solution. Now, note that we cannot have $4\alpha = n - 1$ and $6\gamma = n$ or $4\alpha = n$ and $6\gamma = n - 1$, since we would then have two consecutive even numbers. So we must have $4\alpha = 6\gamma = n - 1$, $5\beta = n$ and $7\delta = n + 1$. This amounts to solving simultaneous congruences, which can be done as follows:

   Since $n - 1 = 4\alpha = 6\gamma$, $n - 1$ must be a multiple of 12. So we set $n = 12n_1 + 1$. Then $\alpha = 3n_1$, $\gamma = 2n_1$. Now $12n_1 + 1 = 5\beta$, so $2n_1 + 1 = 5(\beta - 2n_1)$ must be a multiple of 5. So we set $n_1 = 5n_2 + 2$. Then $n = 60n_2 + 25$ and $\beta = 12n_2 + 5$. Finally, $60n_2 + 26 = 7\delta$, so $4n_2 + 5 - 7(\delta - 8n_2 - 3)$ must be a multiple of 7. So we set $n_2 = 7n_3 + 4$. Then we have $n = 420n_3 + 265$, and a final solution set of

   $$\{a, b, c, d\} = \{2^{105n_3 + 66}, 2^{84n_3 + 53}, 2^{70n_3 + 44}, 2^{60n_3 + 38}\}$$

   for any non-negative integer n_3.

   Alternative Solution:

   A similar approach can yield a slightly easier solution if we instead use the identity $3^m + 3^m + 3^m = 3^{m+1}$.

   First, let $\{a, b, c, d\} = \{3^\alpha, 3^\beta, 3^\gamma, 3^\delta\}$. Then we have $3^{4\alpha} + 3^{5\beta} + 3^{6\gamma} = 3^{7\delta}$. If we can pick α, β, γ and δ such that $4\alpha = 5\beta = 6\gamma = 7\delta - 1 = n$ for some positive integer n, then we have the desired identity.

Since $n = 4\alpha = 5\beta = 6\gamma$, n must be a multiple of 60. So we set $n = 60n_1$. Then we have $60n_1 + 1 = 7\delta$, so $4n_1 + 1 = 7(\delta - 8n_1)$ must be a multiple of 7. So we set $n_1 = 7n_2 + 5$. Then we have $n = 420n_2 + 300$, and a final solution set of

$$\{a, b, c, d\} = \{3^{105n_2+75}, 3^{84n_2+60}, 3^{70n_2+50}, 3^{60n_2+43}\}$$

for any non-negative integer n_2.

7. Evaluate $\sum_{r=1}^{n} 2^r r! + \sum_{j=1}^{n+1} j2^{j+1} j!$.

Solution:

The first step in solving this problem is to combine the two sums (this makes sense as a time-saving device, but is particularly important here since the two cannot be solved separately). We get

$$\sum_{r=1}^{n} 2^r r! + \sum_{j=1}^{n+1} j2^{j+1} j! = \sum_{m=1}^{n} \left[2^m m! + m2^{m+1} m!\right] + (n+1)2^{n+2}(n+1)!.$$

Taking out a common factor of $2^m m!$ in the sum, we get

$$\sum_{m=1}^{n} [(2m+1)2^m m!] + (n+1)2^{n+2}(n+1)!.$$

We would like to find a telescoping form for the summand $(2m+1)2^m m!$ — i.e. an expression of the form $(2m+1)2^m m! = f(m+1) - f(m)$ for some function f. Given the form of the summand, we would expect f to feature both a power of two and a factorial. If we seek a function of this form, it is not hard to find that $f(m) = 2^m m!$ gives

$$f(m+1) - f(m) = 2^{m+1}(m+1)! - 2^m m! = (2m+1)2^m m!$$

as required. So we have

$$\sum_{m=1}^{n}(2m+1)2^m m! = \sum_{m=1}^{n} [f(m+1) - f(m)]$$
$$= f(n+1) - f(1) = 2^{n+1}(n+1)! - 2.$$

Substituting this back into the original sum, we get the final solution:

$$\sum_{r=1}^{n} 2^r r! + \sum_{j=1}^{n+1} j2^{j+1} j! = (n+1)2^{n+2}(n+1)! + 2^{n+1}(n+1)! - 2$$
$$= (2n+3)2^{n+1}(n+1)! - 2$$

8. The volume of a closed right-circular cone of radius r and height h is 1000cm^3. If the radius is increasing at a rate of 3mm per second, at what rate is the total surface area A of the cone changing when $r = 5$cm?

Solution:

The volume of a closed right-circular cone is given by $V = \frac{\pi}{3}r^2 h$, while its surface area is given by $A = \pi r(r + \sqrt{r^2 + h^2})$. If the volume is constant, then we can write the surface area as a function of r alone by isolating h from the first equation and substituting into the second. From the first equation, we get $h = \frac{3V}{\pi r^2}$, and hence $A = \pi r \left(r + \sqrt{r^2 + \frac{9V^2}{\pi^2 r^4}} \right)$. Since we are seeking the rate of change of A with respect to time, and we know the rate of change of r with respect to time, we may use the chain rule:

$$\frac{dA}{dt} = \frac{dA}{dr} \cdot \frac{dr}{dt} = \left[\pi \left(r + \sqrt{r^2 + \frac{9V^2}{\pi^2 r^4}} \right) + \pi r \left(1 + \frac{2r - \frac{36V^2}{\pi^2 r^5}}{2\sqrt{r^2 + \frac{9V^2}{\pi^2 r^4}}} \right) \right] \frac{dr}{dt}.$$

This can be tidied up somewhat to give

$$\frac{dA}{dt} = \left[\frac{2\pi r^3 \left(\pi r^3 + \sqrt{\pi^2 r^6 + 9V^2} \right) - 9V^2}{r^2 \sqrt{\pi^2 r^6 + 9V^2}} \right] \frac{dr}{dt}.$$

At this point, we simply substitute in the given values to find the required value of $\frac{dA}{dt}$. Note that it is important to put all the values in in the same units. So, we have $r = 5$cm, $V = 1000$cm^3 and $\frac{dr}{dt} = 0.3$cm/s, giving

$$\frac{dA}{dt} = -25.0474 \text{ cm}^2/\text{s}.$$

9. Evaluate $\int_0^\pi \frac{x \sin x}{1 + \cos^2 x} dx$.

Solution:

We note that, on the given range, $\frac{\sin x}{1+\cos^2 x}$ is symmetric, so that we have

$$\frac{\sin t}{1 + \cos^2 t} = \frac{\sin(\pi - t)}{1 + \cos^2(\pi - t)}.$$

We can use this symmetry to rewrite the integral over the second half of the range as a similar integral over the first half of the range:

$$\int_{\frac{\pi}{2}}^{\pi} \frac{x \sin x}{1 + \cos^2 x} dx = \int_{0}^{\frac{\pi}{2}} \frac{(\pi - x) \sin(\pi - x)}{1 + \cos^2(\pi - x)} = \int_{0}^{\frac{\pi}{2}} \frac{(\pi - x) \sin x}{1 + \cos^2 x} dx.$$

Using this to simplify the entire integral, we get

$$\int_{0}^{\pi} \frac{x \sin x}{1 + \cos^2 x} dx = \int_{0}^{\frac{\pi}{2}} \left(\frac{x \sin x}{1 + \cos^2 x} + \frac{(\pi - x) \sin x}{1 + \cos^2 x} \right) dx$$
$$= \int_{0}^{\frac{\pi}{2}} \frac{\pi \sin x}{1 + \cos^2 x} dx.$$

Now that the integral is simplified to a purely trigonometric form, we can take a more traditional approach, using a change of variables. Let $u = \cos x$, so that we have $du = -\sin x dx$. Then we get

$$\int_{0}^{\pi} \frac{x \sin x}{1 + \cos^2 x} dx = \pi \int_{0}^{\frac{\pi}{2}} \frac{\sin x}{1 + \cos^2 x} dx = \pi \int_{1}^{0} \frac{-du}{1 + u^2} = \pi \int_{0}^{1} \frac{du}{1 + u^2}.$$

Now, we know that $\int \frac{dt}{1+t^2} = \tan^{-1} t$, so we can substitute this in to get a final result:

$$\int_{0}^{\pi} \frac{x \sin x}{1 + \cos^2 x} dx = \pi \int_{0}^{1} \frac{du}{1 + u^2} = \pi \left(\tan^{-1} 1 - \tan^{-1} 0 \right) = \frac{\pi^2}{4}.$$

Remark:

This problem also appeared as question six on the fourth Superbrain, a full twenty years earlier — so students with long memories were rewarded on this exam!

10. If a, b and c are the lengths of the sides of a triangle, find the least upper bound and greatest lower bound of

$$f(a, b, c) = \frac{a}{b + c} + \frac{b}{c + a} + \frac{c}{a + b},$$

and say if and when these values are attained.

Solution:

Intuitively, we would generally expect the extremes of a function of the sides of a triangle to be attained by the most 'extreme' triangles, that is, the equilateral and degenerate cases. A little experimenting with these

214

should serve to convince the solver that the range of possible values for $f(a, b, c)$ is $[1.5, 2)$. We shall prove each of these results separately.

Part 1: $f(a, b, c) \geq \frac{3}{2}$

We can equip $\frac{a}{b+c}$ with a numerator which is symmetric in a, b and c by adding and subtracting 1. Thus

$$\frac{a}{b+c} = \frac{a+b+c}{b+c} - 1.$$

Applying this to each of the terms, we can then factor out the numerator from each term:

$$f(a, b, c) = (a+b+c)\left(\frac{1}{a+b} + \frac{1}{b+c} + \frac{1}{c+a}\right) - 3.$$

If we then rewrite the first factor in terms of the variables $u = a + b$, $v = b + c$ and $w = c + a$ and rearrange, the inequality we are trying to prove is equivalent to

$$(u+v+w)\left(\frac{1}{u} + \frac{1}{v} + \frac{1}{w}\right) \geq 9.$$

There are a number of ways to prove this inequality. The solver may be familiar with the arithmetic-harmonic mean inequality. This states that for n positive variables $x_1, x_2, \ldots, x_n$, the arithmetic mean is not less than the harmonic mean, i.e.

$$\frac{\sum x_i}{n} \geq \frac{n}{\sum \frac{1}{x_i}}.$$

The desired inequality is simply a rearrangement of the case when $n = 3$. Alternatively, we can apply the arithmetic mean geometric mean inequality to each of the two factors on the left-hand side to get

$$(u+v+w)\left(\frac{1}{u} + \frac{1}{v} + \frac{1}{w}\right) \geq (3\sqrt[3]{uvw})\left(\frac{3}{\sqrt[3]{uvw}}\right) \geq 9.$$

Finally, if the solver is not familiar with either of these techniques, they can multiply out and gather terms. This reduces to

$$\left(\frac{u}{v} + \frac{v}{u}\right) + \left(\frac{v}{w} + \frac{w}{v}\right) + \left(\frac{w}{u} + \frac{u}{w}\right) \geq 6.$$

Now, each of the bracketed expressions is of the form $x + \frac{1}{x}$ for x positive. It is easy to show (by the arithmetic mean geometric mean inequality, or differential calculus, or completing the square, etc) that any such expression must always have a value of at least two, so summing the three gives the desired result.

In any case, this establishes a lower bound for f, and it is easy to see that this bound is achieved if and only if $a = b = c$.

Note that the proof of the first part does not use the fact that a, b and c are the side-lengths of a triangle, so we have proved that

$$\frac{a}{b+c} + \frac{b}{c+a} + \frac{c}{a+b} \geq \frac{3}{2}$$

for all positive real numbers a, b and c. This is known as Nesbitt's inequality.

Part 2: $f(a, b, c) < 2$

This part of the question is (based on people's experience in the exam) a lot harder to do, and in particular is hard to do without first getting an idea of what the upper bound should be. Once we know what to expect, though, the proof is more manageable. If we multiply through on both sides by $(a+b)(b+c)(c+a)$ to clear the denominators and gather terms to one side, the inequality that we need to prove is

$$2(a+b)(b+c)(c+a) - a(a+b)(a+c) - b(b+a)(b+c) - c(c+a)(c+b) > 0.$$

Expanding, we get

$$abc + a^2b + ab^2 + b^2c + bc^2 + ca^2 + ac^2 - a^3 - b^3 - c^3 > 0.$$

Now, if we gather each of the terms that has a factor of a^2 together (and likewise for b^2 and c^2), we get

$$abc + a^2(b+c-a) + b^2(c+a-b) + c^2(a+b-c) > 0.$$

But now, since a, b and c are the side-lengths of a triangle, it is easy to see that each factor in each of the terms on the left-hand side is positive, and so the inequality holds.

Now, it is easy to see that $f(a, b, c)$ can be made arbitrarily close to 2. For example, consider the triangle with side-lengths 1, 1 and $\varepsilon \to 0$:

$$f(1, 1, \varepsilon) = \frac{2}{1+\varepsilon} + \frac{\varepsilon}{2} \to 2.$$

So, 2 is the least upper bound of $f(a, b, c)$, but is never attained.

Twenty-Fifth Superbrain, 2008 — Solutions

1. If A, B and C are the angles of a triangle, prove that
$$\sin A + \sin B > \sin C.$$
Is this result true under the weaker assumption that $A + B + C = \pi$?

Solution:

The inequality given is quite similar to the standard triangle inequality, $a+b > c$, and in fact, this observation is the key to solving the problem.

From the sine rule, we know that the sines of the angles are in proportion to the lengths of the opposite sides: $\sin A = Ka$, $\sin B = Kb$ and $\sin C = Kc$ where a, b and c are the sides opposite angles A, B and C respectively and K is a positive constant. (In fact, $K = \frac{1}{2R}$ where R is the circumradius of the triangle, but it is not necessary to know this.) So, multiplying the triangle inequality through by K, we get
$$\sin A + \sin B = Ka + Kb > Kc = \sin C = \sin(A+B)$$
as required.

If we assume only that $A+B+C = \pi$, then we can have angles outside the open range $(0, \pi)$. Doing this, it is easy to find counterexamples to the claim under this condition, for example $A = 0$, $B = C = \frac{\pi}{2}$ (which corresponds to the equality case of the triangle inequality for a degenerate triangle) gives $\sin A + \sin B = \sin C$, while $A = B = \frac{3\pi}{2}$, $C = -2\pi$ gives $\sin A + \sin B < \sin C$.

Alternative Solution:

If we use the fact that $A + B + C = \pi$, we may rewrite the inequality as
$$\sin A + \sin B > \sin(\pi - A - B).$$
Expanding and simplifying (taking a similar approach to Question 7 from 1993), this can be expressed as
$$4 \sin \frac{A}{2} \sin \frac{B}{2} \sin \frac{A+B}{2} > 0.$$
But now, since A and B are angles of a triangle, A, B and $A + B$ must all lie in $(0, \pi)$, and hence $A/2$, $B/2$ and $(A+B)/2$ must all have positive sines, and the result is clear.

This formulation also makes it easier to systematically find where the inequality does and does not hold under the weaker assumption. In fact, we can easily use this form to partition the plane into the regions where $\sin A + \sin B > \sin C$, $\sin A + \sin B = \sin C$ and $\sin A + \sin B < \sin C$.

2. In how many different orders can ten people sit around a circular table if two particular people, Mister X and Miss Y, must never be seated next to each other?

Solution:

Since there are many ways that two people can be sitting not beside one another, but only one way for them to be sitting beside one another, the easiest way to count the possibilities is to count the number of ways to seat the people without restriction and subtract the number of ways with Mister X and Miss Y seated beside one another.

Since we have a round table, any two positions which differ only by a rotation can be considered the same. So, we may assume that Mister X is in seat number one, with the remaining seats numbered clockwise from there. Since this defines each of the remaining seats uniquely, and since the remaining guests can be seated in any order in these, there are 9! ways to seat the remaining guests.

Now, if we further assume that Mister X and Miss Y must be seated together, then there are only two seats in which Miss Y could sit (given that Mister X is in seat number one — i.e. seats number two and ten). Once this seat is chosen, the remaining guests may be seated in any order in the other eight seats, giving 8! arrangements in each case, and a total of 2.8! arrangements with Mister X and Miss Y seated next to each other.

So, the number of arrangements with Mister X and Miss Y not seated next to one another is simply 9! − 2.8! = 282240.

Alternative Solution:

We can achieve a quicker solution as follows: fix the position in which Mister X sits. Then there are 7 possible positions in which Miss Y may sit (the other nine seats except the two on either side of Mister X). Once those two have been seated, the remaining guests may be seated in any order in the remaining eight seats, which gives 8! different arrangements.

So, we get a total number of arrangements of 7.8! = 282240 as before.

3. If α, β and γ are the solutions of the equation

$$ax^3 + bx^2 + cx + d = 0,$$

and $a \neq 0$, find the value of $\alpha^3 + \beta^3 + \gamma^3$ in terms of a, b, c and d.

Solution:

If α, β and γ are the zeros of the polynomial $ax^3 + bx^2 + cx + d$, then we must have

$$ax^3 + bx^2 + cx + d = a(x - \alpha)(x - \beta)(x - \gamma).$$

Equating the coefficients of powers of x, we get

$$\alpha + \beta + \gamma = -\frac{b}{a} \tag{1}$$

$$\alpha\beta + \beta\gamma + \gamma\alpha = \frac{c}{a} \tag{2}$$

$$\alpha\beta\gamma = -\frac{d}{a} \tag{3}$$

Now, we would like to express $\alpha^3 + \beta^3 + \gamma^3$ in terms of these three functions. It is easy enough to do this by a brute force approach, but we can simplify the calculations somewhat by using the well-known factorization of $x^3 + y^3 + z^3 - 3xyz$ to get

$$\alpha^3 + \beta^3 + \gamma^3 - 3\alpha\beta\gamma = (\alpha + \beta + \gamma)(\alpha^2 + \beta^2 + \gamma^2 - \alpha\beta - \beta\gamma - \gamma\alpha).$$

But now, $\alpha^2 + \beta^2 + \gamma^2 - \alpha\beta - \beta\gamma - \gamma\alpha = (\alpha + \beta + \gamma)^2 - 3(\alpha\beta + \beta\gamma + \gamma\alpha)$, so we can express each factor in terms of the desired functions. Substituting in from (1), (2) and (3), we get

$$\alpha^3 + \beta^3 + \gamma^3 = 3\left(-\frac{d}{a}\right) + \left(-\frac{b}{a}\right)\left(\frac{b^2}{a^2} - \frac{3c}{a}\right) = \frac{3abc - b^3 - 3a^2d}{a^3}.$$

Alternative Solution:

Since α, β and γ are the zeroes of $ax^3 + bx^2 + cx + d$, we have:

$$a\alpha^3 + b\alpha^2 + c\alpha + d = 0, \tag{4}$$

$$a\beta^3 + b\beta^2 + c\beta + d = 0, \tag{5}$$

$$a\gamma^3 + b\gamma^2 + c\gamma + d = 0. \tag{6}$$

Adding the three together and isolating $\alpha^3 + \beta^3 + \gamma^3$, we get

$$\alpha^3 + \beta^3 + \gamma^3 = -\frac{b(\alpha^2 + \beta^2 + \gamma^2) + c(\alpha + \beta + \gamma) + 3d}{a}.$$

But now, substituting in from (1) and (2) above and using the identity $\alpha^2 + \beta^2 + \gamma^2 = (\alpha + \beta + \gamma)^2 - 2(\alpha\beta + \beta\gamma + \gamma\alpha)$, we get

$$\alpha^3 + \beta^3 + \gamma^3 = -\frac{b\left(\frac{b^2}{a^2} - 2\left(\frac{c}{a}\right)\right) + c\left(-\frac{b}{a}\right) + 3d}{a} = \frac{3abc - b^3 - 3a^2d}{a^3},$$

confirming the original result.

Remark:

It is easy to generalize the second solution to give a method for finding $S_n = \alpha^n + \beta^n + \gamma^n$ for $n \geq 3$. If we multiply equation (4) by α^{n-3}, (5) by β^{n-3} and (6) by γ^{n-3}, add the three, and isolate S_n, we get

$$S_n = -\frac{bS_{n-1} + cS_{n-2} + dS_{n-3}}{a}.$$

So we have a recurrence equation for S_n in terms of S_{n-1}, S_{n-2} and S_{n-3}. But we also know the values of the base cases: $S_2 = \frac{b^2 - 2ac}{a^2}$, $S_1 = -\frac{b}{a}$ and $S_0 = 3$. Using this information, we can find S_n in terms of a, b, c and d relatively easily for any n.

4. Evaluate $\displaystyle\int \left(\frac{\sin x}{x}\right)^2 dx - \int \frac{\sin 2x}{x} dx.$

Solution:

Since both integrals are being integrated with respect to x, we can combine the two to get $\displaystyle\int \frac{\sin^2 x - x \sin 2x}{x^2} dx$. The integrand here may remind the solver of the form of the quotient rule:

$$\frac{d}{dx}\left(\frac{u}{v}\right) = \frac{v\frac{du}{dx} - u\frac{dv}{dx}}{v^2},$$

with $v = x$. Indeed, we have $\frac{d}{dx} \sin^2 x = 2 \sin x \cos x = \sin 2x$, so setting $u = \sin^2 x$ and $v = x$ gives

$$\frac{\sin^2 x - x \sin 2x}{x^2} = -\frac{v\frac{du}{dx} - u\frac{dv}{dx}}{v^2} = -\frac{d}{dx}\left(\frac{u}{v}\right) = \frac{d}{dx}\left(-\frac{\sin^2 x}{x}\right).$$

Substituting this into the integral, we get the desired solution easily:

$$\int \left(\frac{\sin x}{x}\right)^2 dx - \int \frac{\sin 2x}{x} dx = -\frac{\sin^2 x}{x} + C,$$

where C is an arbitrary constant.

Alternative Solution:

If the 'inspired' solution is not forthcoming, this problem can be beaten out using power series. The two essential series are

$$\sin x = \sum_{n=0}^{\infty} \frac{(-1)^n x^{2n+1}}{(2n+1)!} \quad \text{and} \quad \cos x = \sum_{n=0}^{\infty} \frac{(-1)^n x^{2n}}{(2n)!}.$$

Noting that $\sin^2 x = \frac{1-\cos 2x}{2}$, we get

$$\left(\frac{\sin x}{x}\right)^2 = \frac{1}{2x^2}\left(1 - \sum_{n=0}^{\infty} \frac{(-1)^n (2x)^{2n}}{(2n)!}\right) = \sum_{n=0}^{\infty} \frac{(-1)^n 2^{2n+1} x^{2n}}{(2n+2)!}. \quad (1)$$

Integrating term-by-term, we get

$$\int \left(\frac{\sin x}{x}\right)^2 dx = \sum_{n=0}^{\infty} \left(\frac{(-1)^n 2^{2n+1}}{(2n+2)!} \int x^{2n} dx\right) = \sum_{n=0}^{\infty} \frac{(-1)^n (2x)^{2n+1}}{(2n+1)(2n+2)!}.$$

Similarly, we get

$$\int \frac{\sin 2x}{x} dx = \sum_{n=0}^{\infty} \left(\frac{(-1)^n 2^{2n+1}}{(2n+1)!} \int x^{2n} dx\right) = \sum_{n=0}^{\infty} \frac{(-1)^n (2x)^{2n+1}}{(2n+1)(2n+1)!}.$$

Subtracting one from the other, we get

$$\int \left(\frac{\sin x}{x}\right)^2 dx - \int \frac{\sin 2x}{x} dx = \sum_{n=0}^{\infty} \left[\frac{(-1)^n (2x)^{2n+1}}{(2n+1)(2n+1)!}\left(\frac{1}{2n+2} - 1\right)\right]$$

$$= -\sum_{n=0}^{\infty} \frac{(-1)^n (2x)^{2n+1}}{(2n+2)!}.$$

Now, note that aside from the minus sign in front and the power of x being larger by one, this is identical to (1) above. So, we conclude that the value of the integral (up to an arbitrary constant) is $-x$ times (1):

$$\int \frac{\sin^2 x - x\sin 2x}{x^2} dx = -x\sum_{n=0}^{\infty} \frac{(-1)^n 2^{2n+1} x^{2n}}{(2n+2)!} + C = -\frac{\sin^2 x}{x} + C.$$

Remark:

Each of the separate integrals given has no solution in terms of elementary functions — so essentially the only way to solve the two was by combining them.

221

5. Find all complex numbers z such that
$$(z+1)^7 = z^7 + 1.$$

Solution:

Expanding the left-hand side and cancelling common terms on both sides, the equation becomes

$$7z^6 + 21z^5 + 35z^4 + 35z^3 + 21z^2 + 7z = 7z(z^5 + 3z^4 + 5z^3 + 5z^2 + 3z + 1) = 0.$$

Now, it is easy to see either from this form or from the original equation that $z = -1$ is a root. Factoring this out, we get

$$7z(z+1)(z^4 + 2z^3 + 3z^2 + 2z + 1) = 0.$$

Now, by the rational root theorem, it is easy to see that $z = \pm 1$ are the only possible rational roots of the quartic term, and it is easy to confirm that neither of these is actually a root. So, we look for quadratic factors. Since we want the first and last terms of the product to have coefficients of one, the factorisation must be of the form

$$z^4 + 2z^3 + 3z^2 + 2z + 1 = (z^2 + az \pm 1)(z^2 + bz \pm 1).$$

Expanding and equating powers of z, we get $a + b = 2$ and $ab = 1$, and hence

$$(x-1)^2 = x^2 - 2x + 1 = x^2 - (a+b)x + ab = (x-a)(x-b).$$

But now, since polynomial factorization is unique, we have $a = b = 1$. So we have

$$z^4 + 2z^3 + 3z^2 + 2z + 1 = (z^2 + z + 1)^2.$$

But now, $z^2 + z + 1$ has the two zeros ω and ω^2, where $\omega = (-1 + i\sqrt{3})/2$ is a complex cube root of unity. So we can factorise $(z+1)^7 - z^7 - 1$ completely into linear factors:

$$(z+1)^7 - z^7 - 1 = 7z(z+1)(z-\omega)^2(z-\omega^2)^2.$$

So, the only roots of this polynomial, and hence the only solutions of the original equation are $z \in \{0, -1, \omega, \omega^2\}$.

6. Evaluate $\lim_{n\to\infty} \left[\frac{1}{n}\sqrt[n]{n!}\right]$.

Solution:

Let $f(n) = \frac{1}{n}\sqrt[n]{n!} = \sqrt[n]{\frac{n!}{n^n}} = \sqrt[n]{\frac{1}{n} \cdot \frac{2}{n} \cdots \frac{n-1}{n} \cdot \frac{n}{n}}$. Then we have

$$\ln f(n) = \frac{1}{n} \sum_{i=1}^{n} \ln\left(\frac{i}{n}\right).$$

But now, for any (Riemann integrable) function g, we have

$$\lim_{n\to\infty} \left[\frac{1}{n} \sum_{i=1}^{n} g\left(\frac{i}{n}\right)\right] = \int_0^1 g(x)\,dx.$$

So we get

$$\lim_{n\to\infty} [\ln f(n)] = \int_0^1 \ln(x)\,dx = \lim_{\varepsilon \to 0^+} \int_\varepsilon^1 \ln(x)\,dx.$$

Since the natural log function has a singularity at zero, we need to evaluate the integral as a limit. Now, $\int \ln(x)\,dx = x\ln x - x$, so we have

$$\lim_{n\to\infty} [\ln f(n)] = \lim_{\varepsilon \to 0^+} [1 \ln 1 - 1 - \varepsilon \ln \varepsilon + \varepsilon] = -1 + \lim_{\varepsilon \to 0^+} [-\varepsilon \ln \varepsilon + \varepsilon].$$

Now, we need to work out the limit of $\varepsilon \ln \varepsilon$, since as ε approaches zero, the two factors approach zero and infinity respectively. Rewriting as

$$\frac{\ln \varepsilon}{\frac{1}{\varepsilon}},$$

we have a ratio of two factors which grow without bound as ε approaches zero, so that we may apply L'Hospital's rule to get

$$\lim_{\varepsilon \to 0^+} \frac{\ln \varepsilon}{\frac{1}{\varepsilon}} = \lim_{\varepsilon \to 0^+} \frac{\frac{1}{\varepsilon}}{-\frac{1}{\varepsilon^2}} = \lim_{\varepsilon \to 0^+} -\varepsilon = 0.$$

Substituting this into the above, we get

$$\lim_{n\to\infty} [\ln f(n)] = \int_0^1 \ln(x)\,dx = -1 + \lim_{\varepsilon \to 0^+} [-\varepsilon \ln \varepsilon + \varepsilon] = -1.$$

Finally, because the natural log function is continuous, we have

$$-1 = \lim_{n\to\infty} [\ln f(n)] = \ln\left(\lim_{n\to\infty} f(n)\right).$$

Taking anti-logs on both sides, we get

$$\lim_{n\to\infty}\left[\frac{1}{n}\sqrt[n]{n!}\right] = e^{-1}.$$

Alternative Solution:

It is known that as n tends to infinity, $\left(1+\frac{1}{n}\right)^n$ increases and tends to e. So, we must have

$$\lim_{n\to\infty} \sqrt[n]{\prod_{k=1}^{n}\left(1+\frac{1}{k}\right)^k} = e.$$

But now, it is easy to see that

$$\prod_{k=1}^{n}\left(1+\frac{1}{k}\right)^k = \frac{(n+1)^n}{n!},$$

and so

$$e = \lim_{n\to\infty}\sqrt[n]{\prod_{k=1}^{n}\left(1+\frac{1}{k}\right)^k} = \lim_{n\to\infty}\left(\frac{n+1}{n}\right)\left(\sqrt[n]{\frac{n^n}{n!}}\right) = \lim_{n\to\infty}\left(n\sqrt[n]{\frac{1}{n!}}\right),$$

from which the result follows.

7. If A and B are distinct points which lie inside a circular disc, show that every point on the line segment $[AB]$ also lies inside the circular disc.

Solution 1:

Let C be a point on $[AB]$, and let $\vec{A}$, $\vec{B}$ and $\vec{C}$ be the vectors from O to A, B and C respectively. Then we must have $|\vec{A}| < 1$, $|\vec{B}| < 1$ and $\vec{C} = t\vec{A} + (1-t)\vec{B}$ for some $t \in [0,1]$.

Now, applying the triangle inequality, we have

$$|\vec{C}| = |t\vec{A}+(1-t)\vec{B}| \leq |t\vec{A}|+|(1-t)\vec{B}| = t|\vec{A}|+(1-t)|\vec{B}| < t+(1-t) = 1,$$

and the result is established.

Solution 2:

Let O be the centre of the disc, and C a point on the line segment $[AB]$. If A and B lie on the same radius, then the result is trivial. So we may assume that $\angle AOB > 0$.

Now, it is known that in any triangle, the largest angle is opposite the longest side and the smallest angle is opposite the smallest side. So, if one side of a triangle is longer than another, then the respective opposite angles are in the same order.

Now, suppose that $|OC| \geq 1$. Then we have $|OC| > |OA|$. So, we must have $\angle OAC > \angle OCA$. Similarly, since $|OC| > |OB|$, we have $\angle OBC > \angle OCB$. Adding the two, we get

$$\angle OAC + \angle OBC > \angle OCA + \angle OCB = \pi.$$

But now, since $\angle OAC = \angle OAB$ and $\angle OBC = \angle OBA$, and since the angle sum of any triangle is π, we have

$$\angle OAB + \angle OBA = \pi - \angle AOB > \pi,$$

which gives $\angle AOB < 0$. But now, since all angles in the plane are non-negative, this is a contradiction. To avoid contradiction, the assumption that $|OC| \geq 1$ must be false. So, we have $|OC| < 1$, and hence C is contained in the disc as required.

Solution 3:

We can obtain a neat solution by noting that a disc is the intersection of half-planes bounded by its tangent lines. So, if A and B lie in a disc, they must lie in each of the tangent half planes. But then, $[AB]$ must lie in each half-plane also (since the line AB and the line bounding any half-plane can have at most one point of intersection, so $[AB]$ cannot go out of the half-plane and then back in), and so $[AB]$ must lie in the disc as required.

8. Given that $\sum_{n=1}^{\infty} \frac{1}{n^2} = \frac{\pi^2}{6}$, evaluate $\sum_{n=1}^{\infty} \frac{1}{n^2(n+1)^2}$.

Solution:

To take advantage of the identity given, we rewrite the numerator of the summand as follows:

$$1 = 1^2 = [(n+1) - n]^2 = (n+1)^2 - 2n(n+1) + n^2.$$

Substituting this in, we get

$$\sum_{n=1}^{\infty} \frac{1}{n^2(n+1)^2} = \sum_{n=1}^{\infty} \frac{1}{n^2} - 2\sum_{n=1}^{\infty} \frac{1}{n(n+1)} + \sum_{n=1}^{\infty} \frac{1}{(n+1)^2}. \quad (1)$$

The first sum is given, and the last sum is simply the first sum except for its first term, so all we need to do to solve the problem is evaluate the simple sum in the middle. Using partial fractions, we get a telescoping sum:

$$\sum_{n=1}^{\infty} \frac{1}{n(n+1)} = \lim_{m \to \infty} \sum_{n=1}^{m} \left[\frac{1}{n} - \frac{1}{n+1}\right] = \lim_{m \to \infty} \left(1 - \frac{1}{m+1}\right) = 1.$$

Substituting this and the given sum into (1), we get

$$\sum_{n=1}^{\infty} \frac{1}{n^2(n+1)^2} = \left(\frac{\pi^2}{6}\right) - 2(1) + \left(\frac{\pi^2}{6} - 1\right) = \frac{\pi^2 - 9}{3}.$$

9. Factorise the expression

$$x^8 + 98x^4 y^4 + y^8$$

as a product of polynomials of degree at least one with integer coefficients which cannot be further factored in the same way.

Solution:

Using the rational root theorem, it is easy to confirm that the given expression has no linear factors with integer coefficients. Moreover, it is not hard to confirm that it also has no quadratic factors with integer coefficients. If we multiply out

$$x^8 + 98x^4 y^4 + y^8 = (x^2 + ax \pm 1)\left(x^6 + \sum_{i=1}^{5} b_i x^i \pm 1\right)$$

and equate coefficients, we get $b_1 = b_5 = -a$, $b_2 = -1 \pm a^2$, $b_4 = a^2 \mp 1$, $b_3 = -a^3 \pm 2a$. Putting all of these in and solving for a, we get $0 = a^4 \mp 4a^2 + 100 = 96 + (a^2 \mp 2)^2$, which clearly has no solutions for real a, and so the expression has no quadratic factors with integer coefficients.

Furthermore, we note that if the expression had a real cubic factor, then there would be a (non-zero, since the cubic cannot have a linear factor) value of $\frac{x}{y}$ which would make the cubic, and hence the entire expression, vanish — since all cubics have a real root. But now, $x^8 + 98x^4 y^4 + y^8 = (x^4 + y^4)^2 + 96(x^2 y^2)^2$, and so the expression does not vanish for any real pair (x, y) except $x = y = 0$. To avoid contradiction, the expression has no real cubic factors. So, if we want to factorize the expression as a product of terms with integer coefficients, it must be factored as a product of two quartics.

We might begin with a naive approach, by looking for a quartic whose square will be the given expression. Clearly such a quartic must contain the terms x^4 and y^4, but these alone will not give a sufficiently large coefficient of $x^4 y^4$ — so, we might add an $x^2 y^2$ term. Then we have

$$(x^4 + ax^2 y^2 + y^4)^2 = x^8 + 2ax^6 y^2 + (a^2 + 2)x^4 y^4 + 2ax^2 y^6 + y^8,$$

which is greater by $x^2 y^2 (2ax^4 + (a^2 - 96)x^2 y^2 + 2ay^4)$ than the given expression. But now, observe that if we can choose a value of a such that this expression is itself a square, then we will have expressed the given polynomial as a difference of two squares, which can be easily factored.

This requires that $2ax^4 + (a^2 - 96)x^2 y^2 + 2ay^4$ must be a square. Since the coefficients of x^4 and y^4 are equal, it must be $2a(x^2 \pm y^2)^2$, and $2a$ must be a square. Equating coefficients, we get $a^2 - 96 = \pm 4a$, which rearranges to $a^2 \pm 4a + 4 = (a \pm 2)^2 = 100 = 10^2$. This gives $a \in \{-12, -8, 8, 12\}$, but $a = 8$ is the only value for which $2a$ is a square. So, setting $a = 8$, we get

$$x^8 + 98x^4 y^4 + y^8 = (x^4 + 8x^2 y^2 + y^4)^2 - \left[4xy(x^2 - y^2)\right]^2,$$

and so the given expression can be completely factorized over the field of polynomials with integer coefficients as

$$(x^4 + 4x^3 y + 8x^2 y^2 - 4xy^3 + y^4)(x^4 - 4x^3 y + 8x^2 y^2 + 4xy^3 + y^4).$$

Alternative Solution:

We can also arrive at this factorization by first going outside the field of polynomials with integer coefficients, and subsequently returning to it.

First, consider the expression as a quadratic in x^4 and using the quadratic formula to get

$$x^4 = \left(-49 \pm \sqrt{49^2 - 1}\right) y^4 = \left(-49 \pm 20\sqrt{6}\right) y^4.$$

Taking square roots on both sides, we get

$$x^2 = \pm \sqrt{-49 \pm 20\sqrt{6}} \; y^2,$$

where the two ± symbols take signs independently of one another. Since the expression under the square root is negative (for either choice of sign), we take out a factor of -1 and look for the (positive) square root of $49 \pm 20\sqrt{6}$.

Let $a + b\sqrt{6} = \sqrt{49 + 20\sqrt{6}}$. Squaring, we get

$$a^2 + 6b^2 + 2ab\sqrt{6} = 49 + 20\sqrt{6}.$$

So, if we equate the coefficients of $\sqrt{6}$ on each side, we get

$$a^2 + 6b^2 = 49, \qquad ab = 10.$$

Substituting $b = a/10$ into the first equation and solving it as a quadratic in a^2, we get $a^2 = 24$ or 25, and hence $a = \pm 2\sqrt{6}$ or ± 5. Using these to get the value of b and substituting into $a + b\sqrt{6}$ (and disregarding negative values, since we are seeking the positive square root), we get $\sqrt{49 + 20\sqrt{6}} = 5 + 2\sqrt{6}$. Similarly, we get $\sqrt{49 - 20\sqrt{6}} = 5 - 2\sqrt{6}$. Substituting these, we get

$$x^2 = \pm i \left(5 \pm 2\sqrt{6}\right) y^2,$$

where, again, the two ± symbols are independent of one another. Taking square roots again, we get

$$x = \pm \sqrt{\pm i (5 \pm 2\sqrt{6})} y.$$

Now, it is easy to confirm that $\sqrt{\pm i} = \frac{1 \pm i}{\sqrt{2}}$. By the same method as above, if we set $c + d\sqrt{6} = \sqrt{5 \pm 2\sqrt{6}}$, square, equate the coefficients of $\sqrt{6}$ and solve the resulting equations, we get $\sqrt{5 \pm 2\sqrt{6}} = \sqrt{2} \pm \sqrt{3}$. Substituting this in, we get

$$x = \pm \left(\frac{1 \pm i}{\sqrt{2}}\right) \left(\sqrt{2} \pm \sqrt{3}\right) y,$$

where all three ± symbols take signs independently of one another. This corresponds to a factorisation of the given polynomial into eight factors of the form

$$x \pm \left[\frac{1}{\sqrt{2}}(1 \pm i)(\sqrt{2} \pm \sqrt{3})\right] y$$

Now, we need to find a way to combine these complex factors into polynomials with integer coefficients. We can begin by multiplying each factor by its complex conjugate, which gives four real quadratic factors of the form $x^2 + \delta_1(2+\delta_2\sqrt{6})xy + (5+2\delta_2\sqrt{6})y^2$, where each of δ_1 and δ_2 can independently assume a value of 1 or -1. Finally, if we fix δ_1 and multiply the two factors with $\delta_2 = 1$ and $\delta_2 = -1$, we get the same two quartic factors with integer coefficients as before:

$$x^4 + 8x^2y^2 + y^4 \pm [4xy(x^2 - y^2)].$$

Remark:

We can find a factorization as a product of two quartic terms by simply expanding $\left(\sum_{i=0}^{4} a_i x^i y^{4-i}\right)\left(\sum_{i=0}^{4} b_i x^i y^{4-i}\right)$, with $a_4 = b_4 = 1$, $a_0 = b_0 = \pm 1$, equating coefficients and solving the resulting equations. However, note that the two proofs above also establish that the two quartic factors are irreducible (i.e. cannot be expressed as a product of polynomials of lower degree with integer coefficients).

10. Place the numbers 1, 2, 3, 4, 5 and 6 in the enclosed boxes in such a way that each digit appears once and only once in each row, each column, each main diagonal and in each of the six outlined sections.

Solution:

To begin with, note that in the top-left 2 × 3 box, the top three boxes cannot contain a 1 (since that row already has one, so the only empty box on the bottom must contain the 1. Similarly, the 2 and 5 in the top-right 2 × 3 box must be in the top row — but the 2 cannot be in the left-hand box since that column already has a 2. So we have the diagram on the right.

Now, in the bottom-right 2×3 box, the 1 can only go in the leftmost box (since the main diagonal and the middle column both already contain 1s), and with this in place the 4 can only go in the rightmost box (since the middle column already contains a 4). Using these, we can fill in the bottom row of the top-right 2 × 3 box. We know that this has to contain 3, 4 and 6 (since 1, 2 and 5 are already in that row). So, the 6 must be in the rightmost box (since that column already contains a 3 and 4), the 3 must be in the middle box (since that column already contains a 4) and the 4 must be in the leftmost box.

			5	1	2
2	1	5	4	3	6
			4		
			2		3
			1		4

We have reached the configuration on the left. Now, we have to place 3 and 6 in the fourth column. Since there is already a 3 on the main diagonal passing through the top box, this must contain the 6 and the 3 must go in the box below. Now, since the rightmost column already contains a 2, the 2 in that 2×3 box must be in the middle box as below right.

Now look at the top-leftmost box. This can only contain a six, since the 2×3 box it is in already has a 1, 2 and 5, while the main diagonal already has a 3 and 4. This means that the 6 in the bottom-right 2×3 box must be in the lower box and the 5 in the box on the main diagonal. Now, since every other number is then on the main diagonal, we can fill a 2 in the last box as shown below left.

			5	1	2
2	1	5	4	3	6
			6	4	
			3	2	
				2	3
				1	4

6			5	1	2
2	1	5	4	3	6
		2	6	4	
			3	2	
			2	5	3
			1	6	4

Now look at the bottom-left 2×3 box. Only the bottom middle box does not share a row or column with a 2, so we fill that in. Then the only empty box not to share a row or a main diagonal with a 3 is the bottom right, so we fill that in. Then the only empty box not to share a row with a 5 is the bottom left, so we fill that in.

Now, the only empty box not to share a column or main diagonal with a 6 is the top right, so we fill that in. Finally, the 1 must go in the top left, since the middle column already has a 1, and the 4 must go in the last box as shown on the right.

6			5	1	2
2	1	5	4	3	6
		2	6	4	
			3	2	
1	4	6	2	5	3
5	2	3	1	6	4

6	3	4	5	1	2
2	1	5	4	3	6
3	5	2	6	4	1
4	6	1	3	2	5
1	4	6	2	5	3
5	2	3	1	6	4

Now it is easy to finish the problem off — first look at the third column (the top box shares a row with a 1, so this must be the 4 and the other the 1), then complete the top-left 2×3 box (which only has one space left) and the rightmost column (where the bottom box now cannot have a 1). Then the last four boxes clearly only have one possibility each. The final solution is shown on the left.